Workshop on Connecting the Dots 2016

EPJ Web of Conferences Volume 127 (2016)

Vienna, Austria
22 - 24 February 2016

Editors:

Rudolf Fruhwirth
Erica Brondolin
Beruadelle Kolbinger
Wolfgang Waltenberger

ISBN: 978-1-5108-3282-4

Printed from e-media with permission by:

Curran Associates, Inc.
57 Morehouse Lane
Red Hook, NY 12571

Some format issues inherent in the e-media version may also appear in this print version.

Printed by Curran Associates, Inc. (2017)

For additional information, please contact EDP Sciences – Web of Conferences
at the address below.

EDP Sciences – Web of Conferences
17, Avenue du Hoggar
Parc d'Activité de Courtabœuf
BP 112
F-91944 Les Ulis Cedex A
France

Phone: +33 (0) 1 69 18 75 75
Fax: +33 (0) 1 69 28 84 91

contact-edps@webofconferences.org

Additional copies of this publication are available from:

Curran Associates, Inc.
57 Morehouse Lane
Red Hook, NY 12571 USA
Phone: 845-758-0400
Fax: 845-758-2633
Email: curran@proceedings.com
Web: www.proceedings.com

TABLE OF CONTENTS

These are the conference proceedings of the workshop "Connecting The Dots 2016" that was held in Vienna, Austria, from February 22 to February 24, 2016, see:

`http://ctd2016.hephy.at`

The workshop concentrated on track reconstruction and other problems in pattern recognition in sparsely sampled data. Such problems are abundant in particle physics, and, consequently, most contributions were rooted in high energy physics experiments and primarily revolved around the reconstruction of particle tracks and jets. Nevertheless, the workshop was intended to be cross-disciplinary; similar problems in other fields were identified and discussed, and contributions from astrophysics, astronomy, or antiproton decelerator facilities also appeared on the agenda.

Talks were by invitation only; in total, 74 participants attended the workshop, 39 talks were given, 16 of which are represented in these conference proceedings. A "Young Scientists Forum" gave our younger colleagues the opportunity to claim the stage in front of an international audience and present their own, personal work. Within this workshop, a meeting of the "Common Track Reconstruction Forum" was held, aiming at an efficient exchange of concepts, algorithms and, ultimately, software:

`http://hepsoftwarefoundation.org/workinggroups/2016/02/02/tracking.html`

The workshop has been received positively by our various scientific communities. It is therefore our goal to establish a "Connecting the Dots" workshop series. Following the inaugural workshop last year at LBNL Berkeley and this year's workshop in Vienna, the series will be continued in March of next year at LAL Orsay, France. Detailed information about the workshop will appear online in due course.

The local organizers:

Rudi Frühwirth, Erica Brondolin, Bernadette Kolbinger, Wolfgang Waltenberger

edp sciences

EPJ Web of Conferences

Statement of Peer review

In submitting conference proceedings to EPJ Web of Conferences, I certify to the Publisher that I adhere to the **Policy on Publishing Integrity** of the journal in order to safeguard good scientific practice in publishing.

1. All articles have been subjected to peer review administered by the proceedings editors.
2. Reviews have been conducted by expert referees, who have been requested to provide unbiased and constructive comments aimed, whenever possible, at improving the work.
3. Proceedings editors have taken all reasonable steps to ensure the quality of the materials they publish and their decision to accept or reject a paper for publication has been based only on the merits of the work and the relevance to the journal.

Title, date and place of the conference

Connecting the Dots 2016, 22-24 Feb. 2016, Vienna, Austria

Proceedings editor(s):

Rudolf FRÜHWIRTH
Wolfgang WALTENBERGER
Erica BRONDOLIN
Bernadette KOLBINGER

Date and editor's signature

Vienna, 20. 9. 2016

17, avenue du Hoggar - PA de Courtabœuf – BP 112 - 91944 Les Ulis Cedex A (France)
Tél. : 33 (0)1 69 18 75 75 - Fax : 33(0)1 69 07 45 17 – www.edpsciences.org

EPJ Web of Conferences **127**, 00002 (2016) DOI: 10.1051/epjconf/201612700002
Connecting The Dots 2016

Performance Requirements for the Phase-2 Tracker Upgrades for ATLAS and CMS

Duccio **Abbaneo**[1,2,a]

[1] *CERN, Geneva, Switzerland*
[2] *For the ATLAS and CMS Collaborations*

Abstract. The High-Luminosity operation of the LHC poses unprecedented challenges for the design of the upgraded trackers of ATLAS [1] and CMS [2]. The stringent requirements imposed by the high particle density and integrated fluence reduce the phase-space of valid technical solutions, inducing both collaborations to design "all-silicon" trackers. On the other hand constraints and requirements coming for the rest of the detector lead to some different choices, especially for the outer trackers. The main requirements for the two tracking systems are reviewed, discussing the implications for the detector designs and layout, and explaining why some of the technical choices remain different in the two experiments. To conclude, some expected performance figures for the two tracking systems are presented.

1 Requirements for the High-Luminosity LHC tracking systems

1.1 Requirements from higher particle rates

A major challenge for the design of the HL-LHC trackers is the target integrated luminosity of 3000 fb^{-1}, which translates to unprecedented requirements in terms of radiation tolerance, particularly difficult to meet for sensors and on-detector electronics. The outer trackers will have to operate with unspoiled performance throughout the high luminosity program, planned for the decade 2026 – 2035, while it is conceivable that the inner parts of the pixel detectors could be replaced during a long shutdown, if the radiation tolerance requirements cannot be fulfilled.

ATLAS and CMS have performed detailed FLUKA [3] simulations to estimate the radiation exposure of the different detector regions, obtaining consistent results (Figure 1). The estimated levels are about one order of magnitude higher compared to the requirements that were used for the design of the existing trackers [4–7], reaching levels of 10^{16} particles per cm^2 in the innermost pixel regions. The particle fluence depends essentially on r (the distance from the beam axis), while the variation with z (the distance from the nominal average collision point, along the beam axis) is very moderate.

When translating such radiation levels into requirements for the design of the tracking subdetectors, it should be noted that the boundary between the outer tracker and the pixel detector is located at a radius of about 350 mm in ATLAS and 200 mm in CMS. Furthermore, in the present tracker upgrade layouts the first pixel layer is located at 39 mm in ATLAS and at 29 mm in CMS, which translates to

[a] e-mail: Duccio.Abbaneo@cern.ch

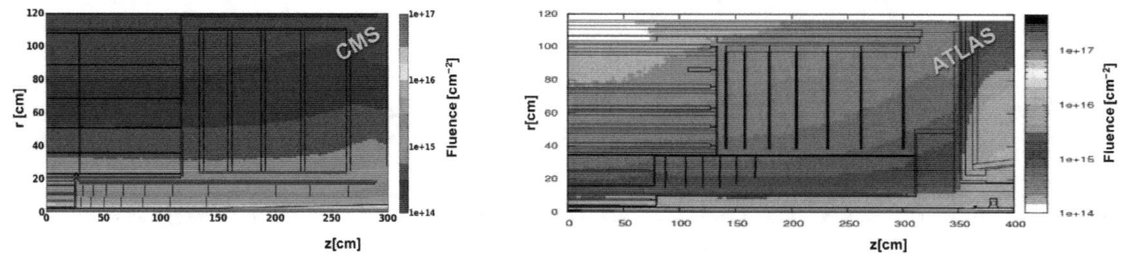

Figure 1. Integrated particle fluence in 1 MeV neutrons equivalent per cm^2, for the CMS [8] and ATLAS [9] phase-2 trackers. The estimates shown correspond to a total luminosity of 3000 fb^{-1} of 14 TeV pp collisions.

nearly a factor of two difference in hit rates and integrated particle fluence per unit area. Some of the reasons for these differences will be discussed below.

The other basic requirement for the high-luminosity operation is the ability to maintain efficient tracking in a high-density particle environment. The experience with the present detectors shows that efficient tracking with low fake rate can be achieved if the channel occupancy does not exceed the 1% level. ATLAS and CMS are designing their trackers to operate with up to 200 pile-up collisions per bunch crossing, which requires a granularity between 5 and 10 times higher than that of the present trackers.

1.2 Requirements from first-level trigger

Besides the fundamental requirements directly linked to the increased particle rates in the tracking detectors, ATLAS and CMS also need to improve their first level trigger decision[1], in order to fully profit from the higher luminosity. The enhancement of the trigger performance involves both a higher output rate of interesting events and an improved discriminating power of the event selection, which is more challenging in high pileup. Improved discriminating power will be achieved by using more information for the trigger decision, with a longer latency available for its processing. ATLAS and CMS plan to enhance the first level trigger rate and latency to the figures shown in Table 1, which are the new requirements for front-end electronics of the tracking systems.

In addition to the increased trigger rate and latency, CMS has chosen to also provide tracking information for the first-level trigger decision, implying that the tracker will send out some information at every bunch crossing. The different design choice is to some extent motivated by different boundary conditions in the two detectors (ATLAS has higher-granularity information from the calorimeters in the first-level trigger, while CMS has a stronger magnetic field, which helps for the specific implementation chosen), and has in turn led to several differences in the design of the upgraded trackers.

Table 1. Planned increase of output rate and latency for the first-level trigger decision in ATLAS and CMS.

ATLAS	100 kHz	⟶	1000 kHz
	2.5 μs	⟶	6.0 μs
CMS	100 kHz	⟶	750 kHz
	3.2 μs	⟶	12.8 μs

[1]N.B.: The first-level trigger is called Level-0 in ATLAS and Level-1 in CMS.

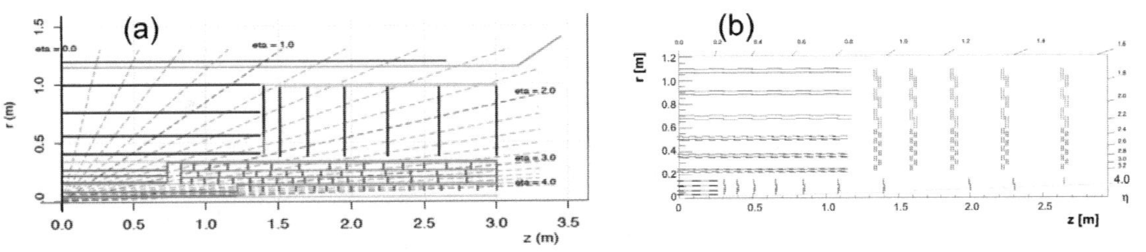

Figure 2. Possible layout of the ATLAS (a) and CMS (b) upgraded tracking systems. In the ATLAS tracker the boundary between outer tracker and pixel detector is located at about 350 mm, while in CMS it is located at 200 mm.

1.3 Other improvements

Both ATLAS and CMS are planning to extend the rapidity acceptance of their tracking systems up to about $|\eta| = 4$. Tracking in the very forward region allows to assign forward jets to primary vertices, which improves performance in vector boson fusion and vector boson scattering physics. The rapidity extension concerns mostly the layout of the pixel detectors, with relevant implications on the detector mechanics and arrangement of the services.

The material in the tracking volume is a major limitation for the performance of the present detectors, affecting not only tracking but also the global event reconstruction. Reducing the amount of material in the upgraded trackers is a vital goal, although the requirements of highly enhanced functionalities mentioned above make such goal challenging to achieve.

ATLAS and CMS are still optimising the layouts of the tracking systems that will have to fulfill the requirements discussed. Two of the layouts under consideration are shown in Figure 2. For the CMS outer tracker, the layout shown is the baseline presented in the upgrade Technical Proposal [8], the collaboration is however also studying a different option with progressively tilted modules in the inner barrel layers. The layouts of the pixel detectors are still in the speculative phase for both collaborations.

2 Outer trackers

2.1 Tracker input to first-level trigger in CMS

The CMS tracker will provide information at each bunch crossing, in order to contribute to the first-level trigger decision. Such functionality relies upon local data reduction in the front-end electronics, in order to limit the volume of data that are sent out at 40 MHz. This is achieved with modules that are capable of rejecting signals from particles below a certain p_T threshold, referred to as "p_T modules" [10–12]. The modules are composed by two closely-spaced sensors read out by a common set of front-end ASICs, that correlate the signals in the two sensors and select the pairs ("stubs") compatible with particles above the chosen p_T threshold (Figure 3). A threshold of around 2 GeV corresponds to a data volume reduction of roughly one order of magnitude, which is sufficient to enable transmission of the stubs at 40 MHz, while all other signals are stored in the front-end pipelines and read out when a trigger signal is received.

The p_T module concept relies on the fact that the strips of both sensors are parallel to the z-axis in the barrel and nearly radial in the end-cap, preventing the use of stereo strips to measure the z

DOI: 10.1051/epjconf/201612700002

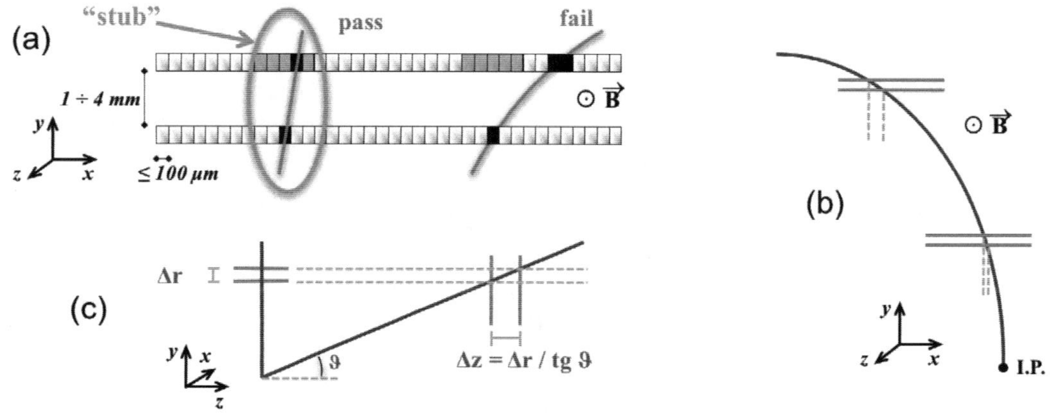

Figure 3. (a) Correlation of signals in closely-spaced sensors enables rejection of low-p_T particles; the channels shown in light green represent the "selection window" to define an accepted "stub". (b) The same transverse momentum corresponds to a larger distance between the two signals at large radii for a given sensor spacing. (c) For the end-cap disks, a larger spacing between the sensors is needed to achieve the same discriminating power as in the barrel at the same radius. The acceptance window can therefore be tuned along with a few different values of sensor spacing to achieve the desired p_T filtering in different regions of the detector. For a realistic pitch of about 100 μm, sufficient p_T resolution can be achieved down to a radius of about 200 mm, thanks to the 3.8 T magnetic field of CMS.

coordinate. For this reason two versions of p_T modules have been realized: 2–Strip (**2S**) modules and Pixel–Strip (**PS**) modules (Figure 4). In PS modules one of the two sensors is segmented into macro-pixels of about 1.5 mm length, providing also a precise measurement of the z coordinate. The main parameters of 2S and PS modules are summarized in Table 2. The choice of pitch of 90 μm and 100 μm are driven by limitations in the line density on the hybrid for the 2S module, and on the bump density of the C4 technology for the PS module. PS modules are deployed in the first three layers of the outer tracker, in the radial region of 200–600 mm (blue in the sketch of Figure 2b), while 2S modules are deployed in the outermost three layers, in the radial region above 600 mm (red in the sketch of Figure 2b). The z coordinates provided by the three PS layers enable some level of primary vertex discrimination for trigger tracks that do not have pixel coordinates, hence the interest to extend the range of PS layers towards low radii, up to the limit where the stub p_T resolution remains acceptable.

Table 2. Main parameters of the 2S module and the PS module of the CMS tracker.

2S module		**PS module**	
~ 2 × 90 cm^2 active area		~ 2 × 45 cm^2 active area	
2 × 1016 strips:	~ 5 cm × 90 μm	2 × 960 strips:	~ 2.5 cm × 100 μm
2 × 1016 strips:	~ 5 cm × 90 μm	32 × 960 pixels:	~ 1.5 mm × 100 μm
Power ~ 5 W		Power ~ 8 W	

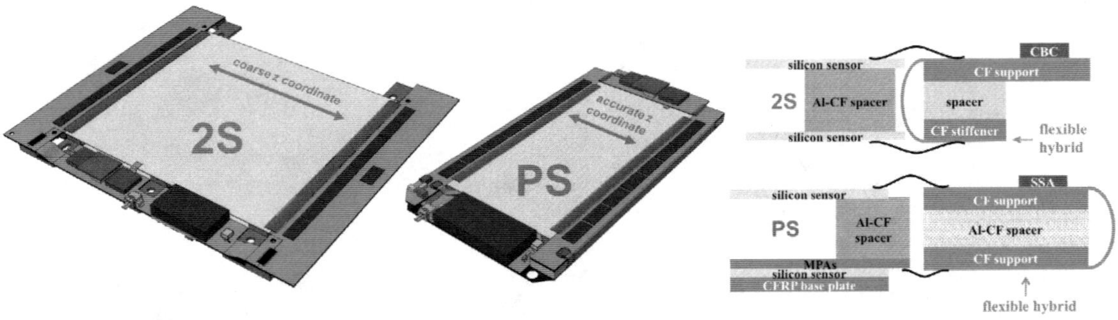

Figure 4. Silicon modules for the CMS tracker. 2S modules are composed of two strip sensors superimposed, PS modules of one strip sensor and one pixelated sensors. In the 2S module, the sensors are read out from the sides, where high-density flex hybrids bring the signals from both sensors to a common set of readout ASICs (CBC, CMS Binary Chip), as shown in the rightmost sketch. In PS modules, one of the two sensors is pixelated, and read out by a dedicated Macro-pixel ASIC (MPA). Also in this case a flex hybrid provides the connectivity between the MPA and the Short-Strip ASIC (SSA), reading out the strip sensor.

2.2 Modules and structures in ATLAS

For the barrel region, ATLAS has designed modules with similar strip lengths as CMS (48 mm and 24 mm for the outer and inner layers, respectively), but could afford a smaller pitch of 75 μm, since the readout electronics collects signals from one sensor only. The readout hybrid is located on top of the sensor (Figure 5a), which avoids large inactive area at the edge of the modules, that can therefore be abutted on the same surface with minimal inefficiency in between.

Barrel layers are composed of "staves" (Figure 5c), each carrying two rows of modules on the two sides. Top and bottom modules have the strips tilted by ± 20 mrad with respect to the z-axis, hence providing also a precise measurement of the z coordinate.

In the end-caps, modules are arranged on "petals" (Figure 5b) that follow the same concept as the barrel staves. Several different sensor designs and variants of hybrid circuits are needed to cover the wedge-shaped surface with abutted modules. The granularity is similar to the barrel, with a typical pitch of 75–80 μm and strip lengths of 20–30 mm in the inner rings, and 40–60 mm in the outer rings. On one side of the petal modules have radial strips, providing the ϕ coordinate, while on the other side the strips are tilted by 40 mrad, hence measuring also r.

2.3 Modules and structures in CMS

The p_T modules discussed in Section 2.1 populate the entire volume of the CMS outer tracker. The two halves of each module are read out independently by the hybrids on the two sides, which prevents the reconstruction of stubs when particles cross the module near the center, with a large incident angle (see sketch of Figure 6b). In a flat barrel layout (Figure 6a) such an effect translates into a geometrical inefficiency of stub finding which is larger than 30% at the edge of the first barrel layer. To overcome this limitation, CMS is studying an alternative layout where the first three barrel layers, that are populated by PS modules, feature progressively tilted modules, nearly perpendicular to incident particles over the entire barrel length (Figure 6c). The design of the mechanical structure for the tilted section is ongoing (Figure 6d) and the option appears to be promising.

Figure 5. ATLAS silicon modules feature the readout hybrid integrated on top of the sensors, hence avoiding inactive area on the sides (a). In the barrel staves (c), two rows of modules are abutted on each side of the mechanical structure. A similar concept is implemented in the end-cap petals, where different sensor and hybrid versions are needed to cover the wedge-shaped surface.

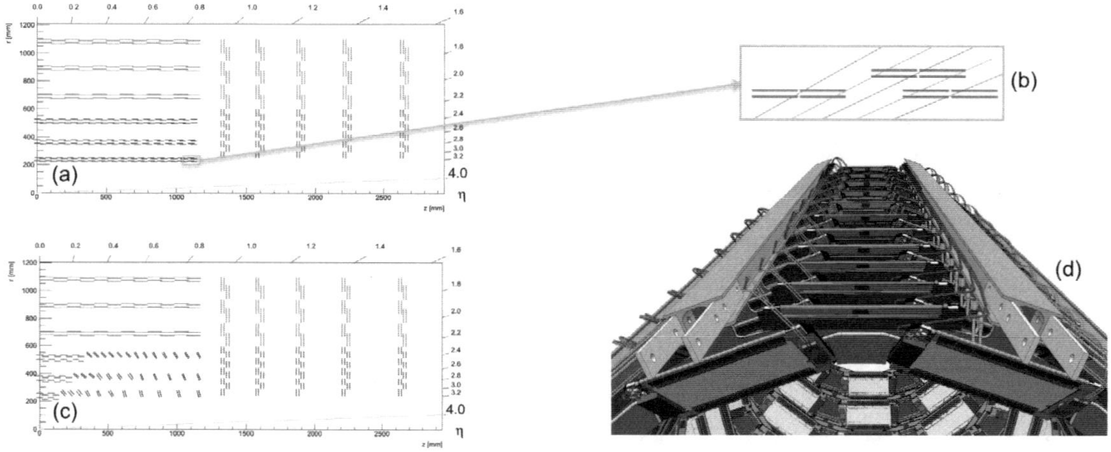

Figure 6. In the flat barrel layout (a) PS modules at the edge of the inner barrel layers have a large geometrical inefficiency for on-board stub finding (b): while the inefficiency at the edge of the modules can be compensated by implementing large overlaps between consecutive modules in z, the inefficiency in the center is irrecoverable (indicated by the red tracks). To overcome the problem, an alternative layout with progressively tilted modules (c) is under study. The design of the mechanics for the tilted layout is well advanced (d).

In the outer three layers of the barrel the effect of stub finding inefficiency is much less severe because the incidence angle of particles is less pronounced , the sensor spacing at those radii is smaller, and the modules are twice as long on the z side. Modules are mounted on "ladders", where overlap in the z view is achieved by placing the modules at different radii (Figure 7a). Along ϕ, consecutive ladders are again staggered in r, hence a hermetic barrel layer is formed by modules located at four different radii.

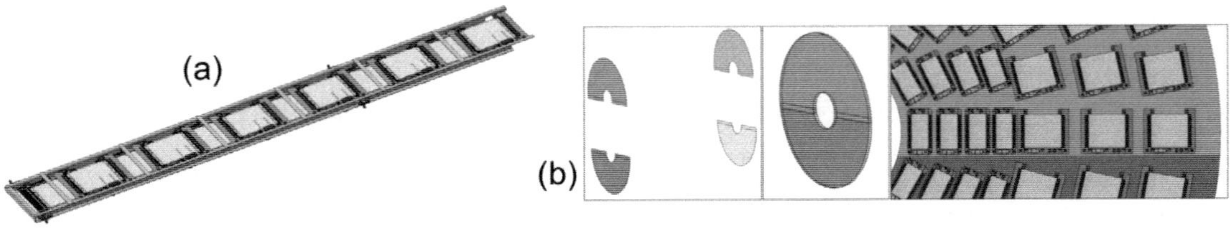

Figure 7. In the ladders of the CMS outer barrel (a), modules are mounted at two different radii in order to implement z overlap. In the end-caps, the same rectangular modules are arranged on the four surfaces of a double-disk structure (b).

In the end-cap, due to the large radial extension (from 200 mm to nearly 1100 mm) and the small size of PS modules, the use of wedge-shaped modules would lead to an impractical large number of sensor and hybrid varieties. Therefore the same rectangular modules used in the barrel are arranged on a double-disk structure: modules located at a given radius are mounted alternately on the two sides of a disk with overlap in ϕ, forming a hermetic ring; consecutive rings along r are mounted on different disks, hence also achieving radial overlap. Similarly to the barrel, a hermetic disk surface is achieved with modules mounted at four different z locations.

3 Pixel detectors

The high-luminosity operation implies extreme challenges for the design of the pixel detectors in terms of radiation tolerance of sensors and readout electronics, as well as data volume to be stored in the front-end pipelines and sent out at high trigger rates. In order to cope with the reduced charge in the sensors, a readout chip with small cell size and low detector threshold is required. ATLAS and CMS are carrying out a common development in the framework of RD53 [13], to design a pixel chip with 2500 μm^2 cell size, in 65 nm CMOS technology.

Despite the common development for the front-end ASIC, some relevant differences remain in the layout of the two pixel detectors, in their present stage of development. Besides the different outer boundary discussed above, motivated by the fact that CMS implements macro-pixels in the intermediate radial range, in the current designs the inner boundary is also different, with the CMS pixel detector featuring a first layer at 29 mm average radius, while the ATLAS detector has the first layer at 39 mm. The difference is quite relevant as it translates to nearly a factor of two difference in hit rates and particle fluence per unit surface. The difference is partially justified by the fact that the CMS pixel detetctor can be installed and extracted with the beam pipe in place, and is removed in the occasion of a beam pipe bakeout; hence it can be located as close as allowed by the mechanical tolerances. In comparison, in ATLAS the first layer of the pixel detector is mounted directly onto the beam pipe, which requires a certain thickness of insulation material for the beam pipe bakeout. CMS maintains as its design target to have a first layer at about 30 mm also for the high luminosity operation, while ATLAS is exploring the possibility of reducing the radius compared to the current design.

Another topic under study is the optimal pixel aspect ratio, for the assumed pixel surface of 2500 μm^2. CMS has studied the case of the most challenging barrel layer 1, comparing the $r - \phi$ and z resolutions obtained on the single hit, for different assumptions on sensors thickness and chip detection threshold. The results, summarized in the plots of Figure 8, indicate that, notably in the

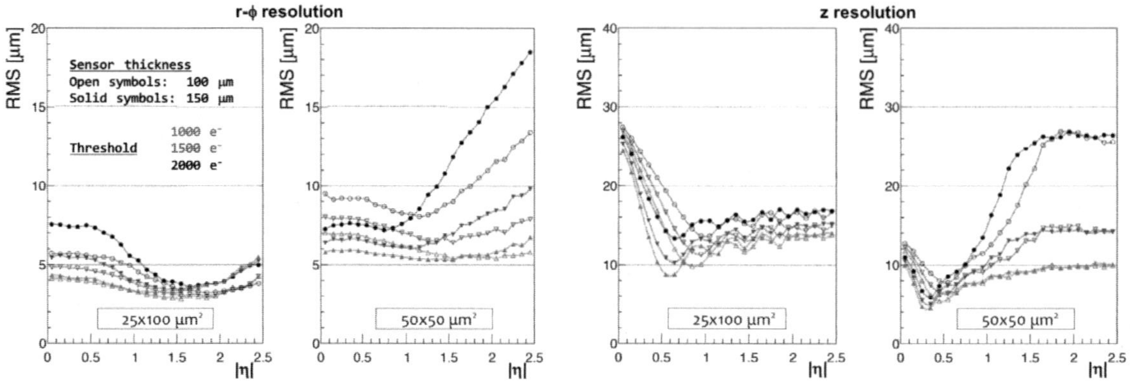

Figure 8. Comparison of single hit resolutions in $r-\phi$ and z obtained in the first pixel barrel layer with square and rectangular pixels, for different assumptions on sensor thickness and detection threshold. Open and solid markers correspond to 100 μm and 150 μm silicon thickness, respectively; red markers correspond to a threshold of 1000 e^-, blue markers 1500 e^- and black markers 2000 e^-. Square pixels offer better z resolution in the central region, while rectangular pixels give in general better performance in the other cases, and are less demanding on the chip threshold. These results are obtained with the flat barrel geometry of Figure 2b, and the simulation corresponds to unirradiated sensors.

forward part of the barrel, rectangular pixels of $25 \times 100\ \mu m^2$ offer better resolution, while being less demanding in terms of detection threshold and required bandwidth (which is particularly relevant for a first layer located at a radius of 30 mm). ATLAS instead is focussing on the use of square pixels studying how to exploit the correlation between cluster length and track incident angle: details on these studies can be found in [14].

Concerning the overall layout, CMS has so far used the geometry shown in Figure 2b, derived from the "phase-1" detector with added disks to extend the rapidity coverage. It is however known that modifications to this geometry will be needed, because the detector insertion mechanism requires a larger opening in the forward part, beyond $z \approx 1500$ mm. The boundary between pixel detector and outer tracker will have to be moved to a higher radius in that region.

ATLAS has several pixel detector concepts under study. As an alternative to the flat barrel layout, two different options implementing progressively tilted modules are investigated, and could be deployed in some or all of the layers, as shown in the sketches of Figure 9. The design of both versions of the tilted geometry is well advanced, with realistic prototypes already built (Figure 10).

4 Expected performance

Both collaborations have produced performance estimates for the tracker concepts under study. In Figure 11 the efficiency as a function of transverse momentum is shown for the ATLAS tracker, along with the muon p_T resolution and the fraction of mis-reconstructed tracks. The solid black markers correspond to the baseline layout, offering good performance up to $|\eta| < 4$, while the other markers correspond to cost-saving options with reduced coverage.

In Figure 12 the expected performance of the CMS upgraded tracker at 140 pileup is compared to the "phase-1" detector at 50 pileup. The upgraded tracker provides equivalent or better performance in the higher pileup conditions within the common acceptance range, while extending the acceptance

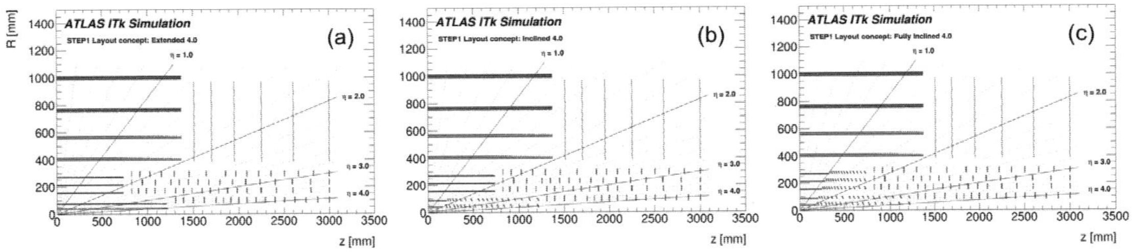

Figure 9. Layout of the ATLAS tracker with three different options for the pixel barrel: a geometry with flat layers (a), an option where the first two layers implement progressively tilted modules (b), and a third version with tilted geometry in all layers (c).

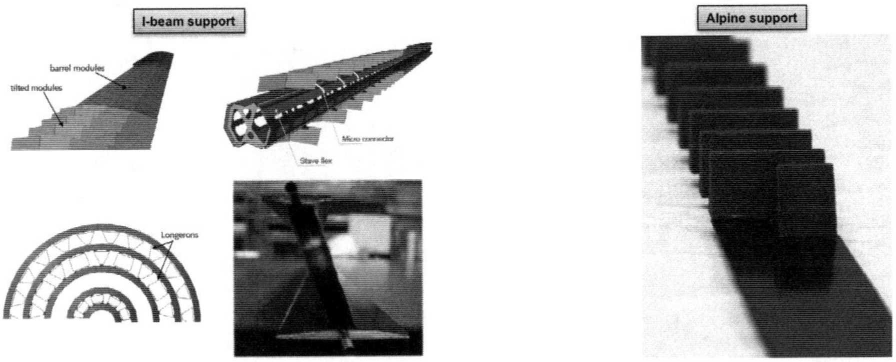

Figure 10. Two different options are being developed in ATLAS [15] to implement tilted modules in pixel barrel layers. In the I-beam concept (left) mechanical supports are shared between neighbouring layers, offering the possibility to realize lightweight structures of high stiffness, particularly suitable for long layers. The Alpine concept (right) is a straightforward adaptation of the flat stave concept, where carbon-fiber wedges are added to support modules with the desired tilt angle.

up to nearly $|\eta| = 4$. The amount of material of the outer tracker is reduced by about a factor of two compared to the present tracker, despite the highly enhanced functionalities.

Another important aspect for the CMS tracker is the performance in the reconstruction of tracks for the first-level trigger, which was the subject of other contributions to this workshop [16].

5 Conclusions

The high-luminosity operation of the LHC entails unprecedented requirements in terms of particle density and integrated fluence for the trackers of ATLAS and CMS. Such requirements limit the phase-space of valid technical solutions, inducing both Collaborations to design "all-silicon" trackers.

The challenge is particularly difficult for the pixel detectors, where ATLAS and CMS are carrying out a common development for the readout chip, in the framework of RD53.

CMS has chosen to implement tracking in the first-level trigger decision, which has driven several specific design choices, especially in the outer tracker: the implementation of p_T modules limits the achievable pitch and requires the development of macro-pixels to measure the z coordinate; the

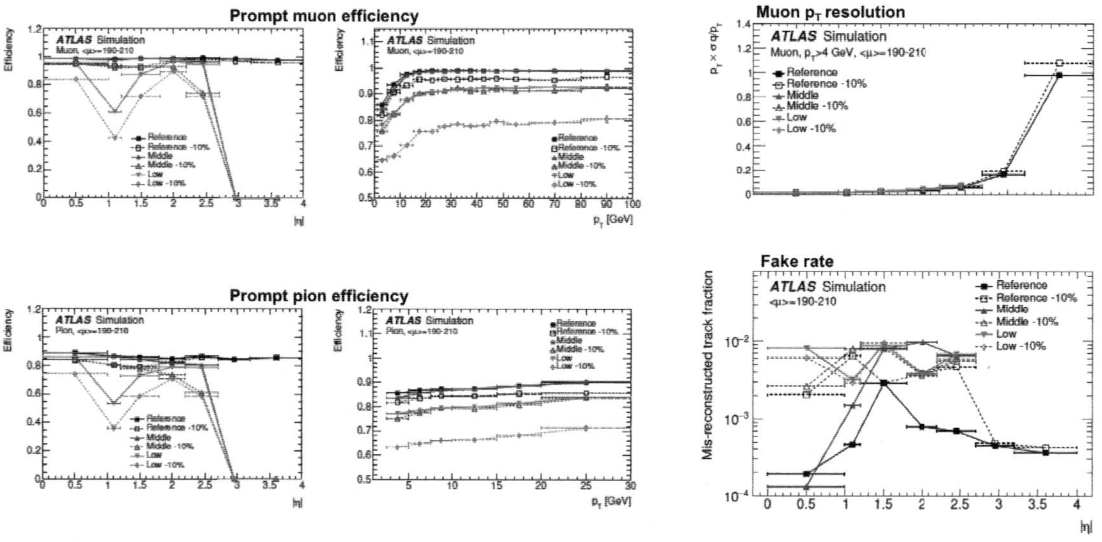

Figure 11. Main performance figures for the upgraded ATLAS tracker [15]. The baseline layout, corresponding to the solid black markers, provides excellent tracking performance up to $|\eta| = 4$. The other markers correspond to various cost-saving options with reduced coverage.

boundary between outer tracker and pixel detector is pushed to lower radii to achieve primary vertex discrimination in trigger tracks; the large radial extension and half-size PS module disfavour the development of wedge-shaped modules for the end-cap; the need of mitigating the geometrical inefficiency for stub finding has motivated the development of the tilted layout for the inner barrel. The sketch of Figure 13 illustrates the cascade of implications.

The available performance estimates indicate that both detectors will provide efficient tracking in high pileup conditions, and at the same time offer improved resolution and extended acceptance.

While the design of the outer trackers is relatively advanced, layout options for the pixel detectors are still being explored, and further developments are underway.

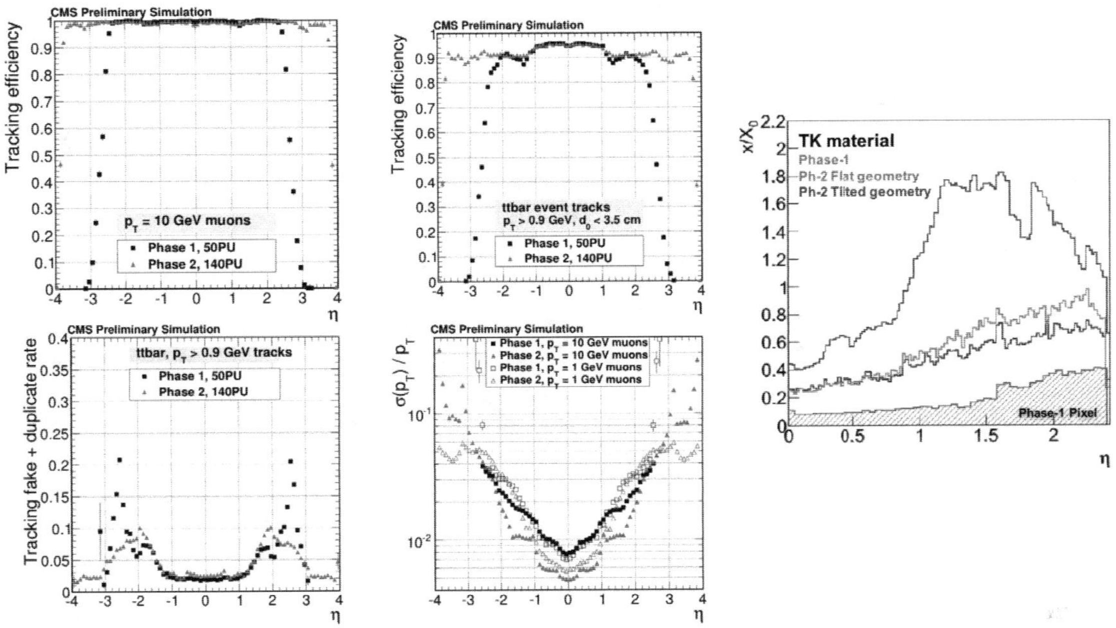

Figure 12. The performance of the upgraded CMS tracker in 140 pileup events is compared to the phase-1 tracker in 50 pileup events [8]. The upgraded tracker provides similar tracking efficiency and better resolution and fake rate, plus extended acceptance. The amount of material in the outer tracker is substantially reduced, and the tilted geometry for the inner barrel offers an additional reduction compared to the flat geometry. N.B.: A material estimate for the upgraded pixel detector is not yet available, therefore the phase-1 material distribution is used in this plot.

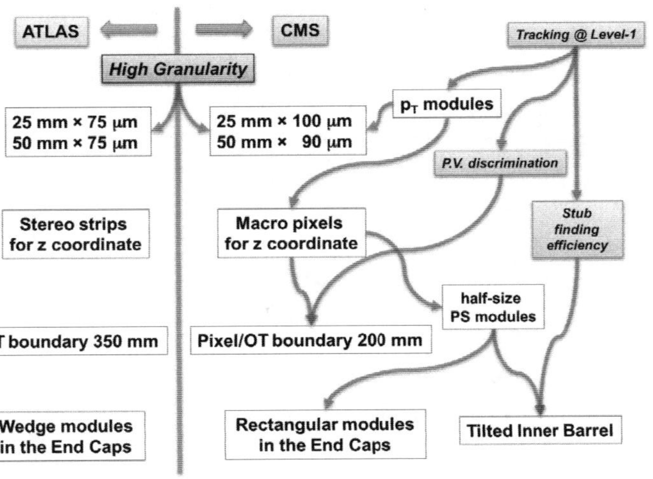

Figure 13. The sketch illustrates how the requirement of contributing tracking information to the first-level trigger decision for the CMS tracker has resulted in a cascade of different technical choices compared to the ATLAS tracker.

References

[1] The ATLAS Collaboration, JINST **3**, S08003 (2008), http://stacks.iop.org/1748-0221/3/i=08/a=S08003
[2] The CMS Collaboration, JINST **3**, S08004 (2008), http://stacks.iop.org/1748-0221/3/i=08/a=S08004
[3] A. Ferrari, P.R. Sala, A. Fassò, J. Ranft, Tech. Rep. CERN-2005-010, CERN, Geneva, Switzerland (2005), https://cds.cern.ch/record/898301
[4] The CMS Collaboration, Tech. Rep. CERN-LHCC-1998-006, CERN, Geneva, Switzerland (1998), https://cds.cern.ch/record/368412
[5] The CMS Collaboration, Tech. Rep. CERN-LHCC-2000-016, CERN, Geneva, Switzerland (2000), https://cds.cern.ch/record/490194
[6] The ATLAS Collaboration, Tech. Rep. CERN-LHCC-1997-016, CERN, Geneva, Switzerland (1997), https://cds.cern.ch/record/331063
[7] The ATLAS Collaboration, Tech. Rep. CERN-LHCC-1997-017, CERN, Geneva, Switzerland (1997), https://cds.cern.ch/record/331064
[8] The CMS Collaboration, Tech. Rep. CERN-LHCC-2015-010, CERN, Geneva, Switzerland (2015), https://cds.cern.ch/record/2020886
[9] The ATLAS Collaboration, Tech. Rep. CERN-LHCC-2012-022, CERN, Geneva, Switzerland (2012), https://cds.cern.ch/record/1502664
[10] C. Foudas, A. Rose, J. Jones, G. Hall (2005), https://arxiv.org/abs/physics/0510227
[11] G. Hall, M. Raymond, A. Rose, JINST **5**, C07012 (2010), http://iopscience.iop.org/article/10.1088/1748-0221/9/10/C10034/meta
[12] M. Pesaresi, G. Hall, JINST **5**, C08003 (2010), http://iopscience.iop.org/article/10.1088/1748-0221/5/08/C08003/meta
[13] J. Christiansen, M.L. Garcia-Sciveres (eds.), Tech. Rep. CERN-LHCC-2013-008, CERN, Geneva, Switzerland (2013), https://cds.cern.ch/record/1553467
[14] S. Viel et al., Nucl. Inst. & Meth. A **831**, 254 (2016), http://www.sciencedirect.com/science/article/pii/S0168900216301346
[15] The ATLAS Collaboration, Tech. Rep. CERN-LHCC-2015-020, CERN, Geneva, Switzerland (2015), https://cds.cern.ch/record/2055248
[16] G. Fedi, these proceedings; L. Skinnari, these proceedings

EPJ Web of Conferences **127**, 00003 (2016) DOI: 10.1051/epjconf/201612700003
Connecting The Dots 2016

4D Cellular Automaton Track Finder in the CBM Experiment

Valentina **Akishina**[1,2,a] and Ivan **Kisel**[1,2,b]

[1] *Goethe University, Frankfurt am Main, Germany*
[2] *Frankfurt Institute for Advanced Studies (FIAS), Frankfurt am Main, Germany*

Abstract. The CBM experiment (FAIR/GSI, Darmstadt, Germany) will focus on the measurement of rare probes at interaction rates up to 10 MHz with data flow of up to 1 TB/s. It requires a novel read-out and data-acquisition concept with self-triggered electronics and free-streaming data. In this case resolving different collisions is a non-trivial task and event building must be performed in software online. That requires full online event reconstruction and selection not only in space, but also in time, so-called 4D event building and selection. This is a task of the First-Level Event Selection (FLES).
The FLES reconstruction and selection package consists of several modules: track finding, track fitting, short-lived particles finding, event building and event selection. The Cellular Automaton (CA) track finder algorithm was adapted towards time-based reconstruction. In this article, we describe in detail the modification done to the algorithm, as well as the performance of the developed time-based CA approach.

1 Introduction

The CBM experiment [1] is targeted to explore the QCD phase diagram in the region of high baryon densities by investigating nuclear collisions from 2 to 45 *A*GeV (per nucleon) beam energy. The CBM detector is designed to measure rare diagnostic probes such as multi-strange hyperons, charmed particles and vector mesons decaying into lepton pairs with unprecedented precision and statistics. In order to perform measurements with the required precision, the experiment is being designed to operate under interaction rates of up to 10 MHz, which puts strong constraints on the data processing. Together with the high multiplicity of charged particles produced in heavy-ion collisions — up to 1000 particles in central collisions — this leads to huge data rates of up to 1 TB/s. The data acquisition is thus being designed in a free-running fashion, without a hardware trigger. Event reconstruction and selection will be performed online on a dedicated many-core CPU/GPU computer farm.

The traditional latency-limited trigger architectures, typical for conventional experiments with a hardware trigger, are not suited to the case of CBM. Instead, the experiment will ship and collect time-stamped data into a readout buffer in the form of a time-slice of a certain time length, not in the form of isolated collisions, and deliver it to a large computer farm, where online event reconstruction and selection will be performed. In this case a considerable fraction of collisions inside a time-slice may potentially overlap in time with each other. The reconstruction algorithms in this case should be

[a] e-mail: V.Akishina@gsi.de
[b] e-mail: I.Kisel@compeng.uni-frankfurt.de

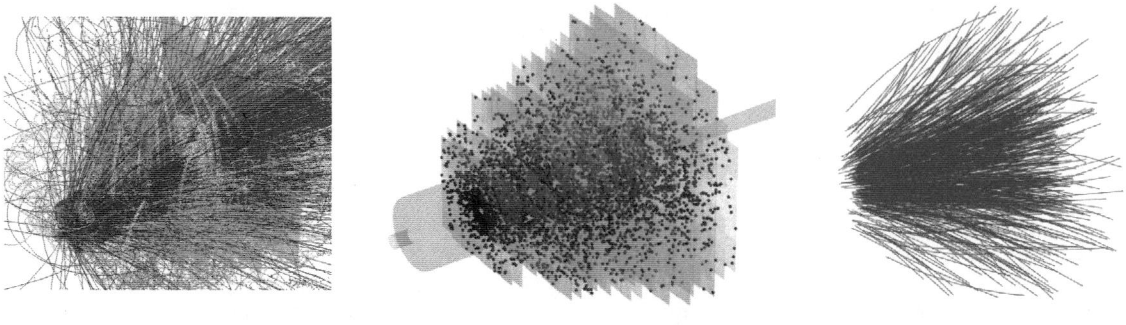

Figure 1. An illustration of the complexity of the track finding problem. From left to right: particle trajectories in a simulated central Au+Au collision at 25 *A*GeV energy in the CBM experiment; only hits in the main tracking detector as the input information for track finding; the reconstructed tracks found by the CA track finder. For the simulation of the events the URQMD generator was used in conjunction with GEANT3 for the transport of the generated particles through the detector setup.

modified in order to process not event-associated data. Association of hit information with physical events must be performed in software and requires full online event reconstruction not only in space, but also in time, so-called 4-dimensional track reconstruction.

2 Cellular automaton based track finder

Since one of the most challenging and time-consuming part of collision reconstruction is track finding, optimization of a track finder algorithm has great impact on the computational effort needed for the data-analysis chain (see Figure 1). The reason for that is that the track finder usually takes as an input raw detector hit measurements at the very first event reconstruction phase, when no data reduction can be done yet. Therefore one of the major obstacles to be solved by each track finder is a huge and fast growing with track density amount of combinations, which the algorithm has to consider in order to bind together one- or two-dimensional measurements into five-dimensional tracks.

The exponential growth of the combinatorial enumeration at high track densities usually makes it impossible to consider all combinations within a reasonable time. A solid solution for combinatorial optimization is provided by the Cellular Automaton (CA) approach. The CA track finder features make the algorithm an appropriate solution for track reconstruction in the main tracking detector of CBM, the Silicon Tracking System (STS).

The CA method profits from building up so-called cells, i.e. short track segments (triplets in the case of CBM) with a higher dimensionality than measurements, before starting the main combinatorial search. These cells are combined into reconstructed tracks during the CA track finder evolution — the process of defining neighboring cells, which with high probability belong to the same track. The rules of CA evolution in this case are given by the definition of the neighboring segments, e.g. in CBM that is the triplets, which coincide in the position and momentum within a certain precision.

The algorithm is intrinsically local, working with data only in a neighborhood. It provides a steady consolidation of the track information identified in the course of the algorithm evolution. In addition, the CA based algorithms can be parallelized in order to be implemented on many-core CPU/GPU computer architectures. Let us consider some details of the CBM CA track finder algorithm.

2.1 CA track finder algorithm strategy

A charged particle track in a magnetic field has five degrees of freedom. Hence, a particle trajectory can be fully described by five parameters. The CBM state vector was chosen in the following form: $(x, y, t_x, t_y, q/p)$. Here, x, y are the track coordinates at a particular z position; t_x, t_y are the slopes of the track with respect to the z-axis; and q/p is the particle charge to momentum ratio.

Let us define a triplet as a group of three hits on adjacent detector stations, potentially produced by the same particle. Since a hit consists of two position measurements (x_i, y_i) of the intersection of a particle with a certain detector plane, the triplet represents a set of six coordinate measurements and allows to uniquely determine all five parameters of a reconstructed track segment. For this reason in CBM the short track segments to be built at the first stage are triplets.

The parameters used to describe a triplet are the same as the track parameters: $(x, y, t_x, t_y, q/p)$. Thus, the Kalman filter equations [2] for the estimate of the χ^2-deviation between the hit measurements and the parameters of the reconstructed track segments are the same as for the full track. It gives the opportunity to reject according to the χ^2-value most of the random three-hit combinations already at the triplet level.

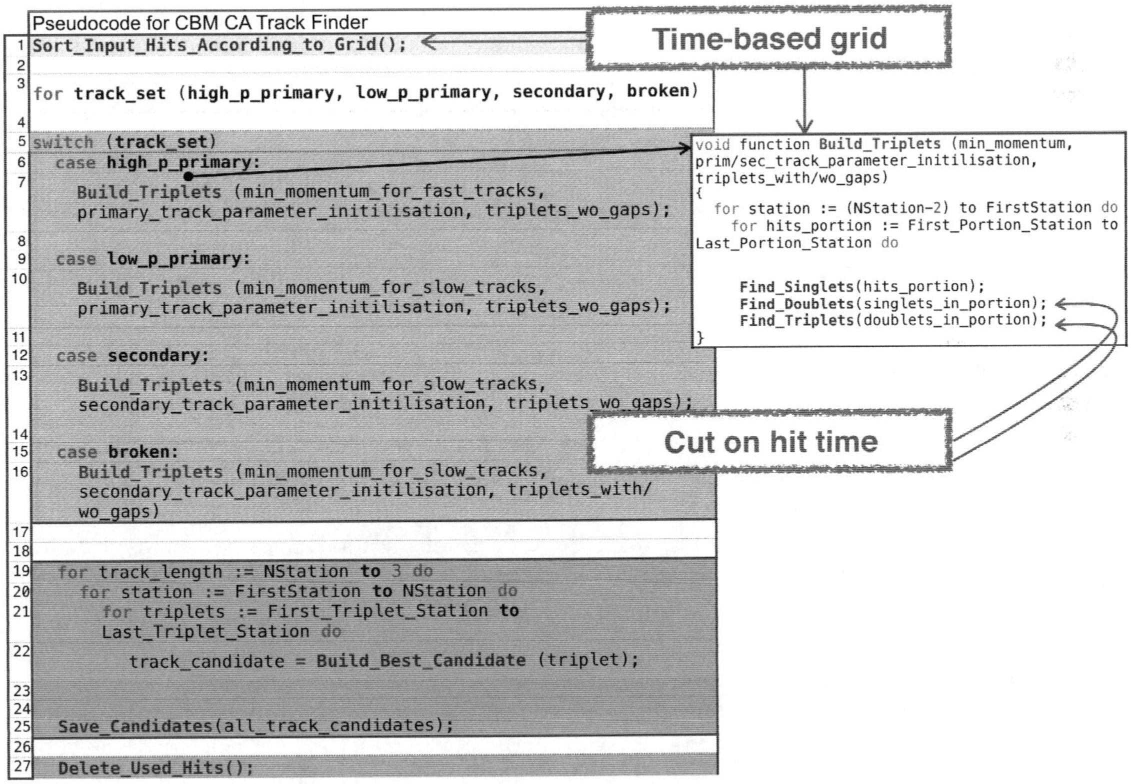

Figure 2. A pseudocode scheme for the CA track finder algorithm.

The strategy of the algorithm can be explained with the help of the pseudocode scheme that is shown in Figure 2. The tracking procedure starts with the initialization (Figure 2: pseudocode line 1). At this stage the algorithm allocates memory for the input information and prepares a grid data structure to store hits (Figure 3).

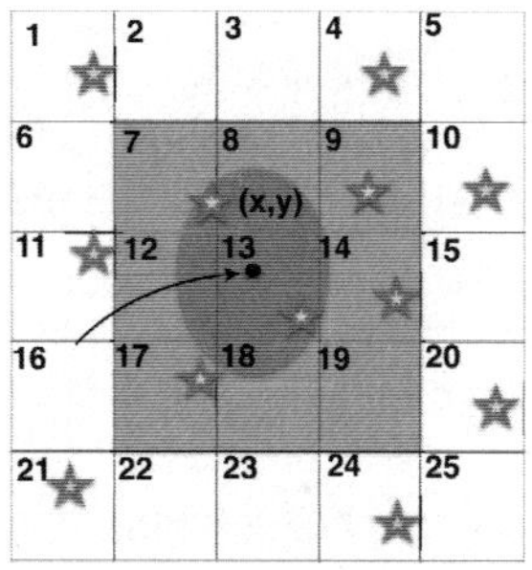

Figure 3. The grid structure provides fast access towards hit measurements in the area of track extrapolation within the extrapolation errors.

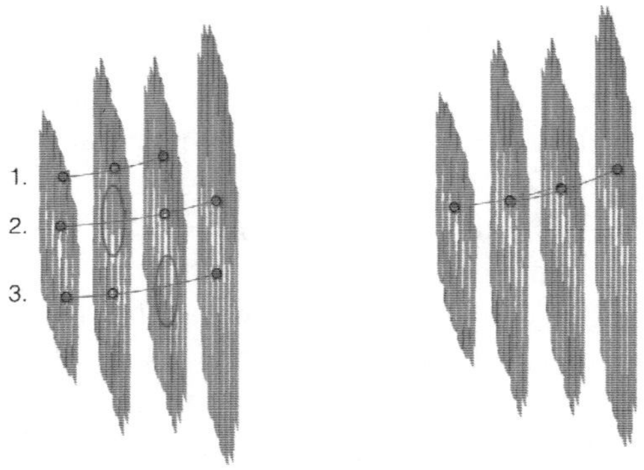

Figure 4. Left: three types of triplets built by the CA algorithm: 1) with no missing hits, 2) with the second hit missing, 3) with the third hit missing. Right: two neighboring triplets are combined into a track.

The grid structure plays an important role for the speed of the algorithm, since it provides fast access to hits for the most time-consuming part of triplets construction. While binding hits into triplets, one often faces the task of finding a hit of a triplet in a certain spatial region of a station. In order to do it fast, a regular 2D grid is introduced in each station, with the grid cell size inversely proportional to the hit density on the station. All hits are sorted and stored in the memory in the order of their grid cell numbers. For each grid cell a pointer to the first hit is stored. After extrapolation of the track segment to the next station, the algorithm obtains an estimated position (x, y) with extrapolation errors. It defines a certain region of interest on the station, where potentially a next hit can be located. One can use the grid structure to avoid going through the entire list of hits on the station and to check whether they lie in the region of interest.

After determining the region of interest, the algorithm builds a special object called HitArea, which is defined by the area of bins fully covering the region of interest (shown by the blue square in Figure 3). The grid structure allows a fast access to the hits lying within the HitArea. Since the grid stores indexes of the first hits in the bin for each bin, to check all the hits in the HitArea, the algorithm just needs to check several groups of hits, sorted and located locally in the memory, instead of going through the whole list of hits on the station.

The most time-consuming part of the algorithm (90.4% of the total time) is the triplet building stage. Triplets are built out of hit measurements that can potentially belong to the same track. Since the tracking algorithm is designed for online reconstruction, the algorithm speed plays a crucial role and a lot of optimization efforts were devoted to speed up the execution. One such optimization comes into play at the very beginning.

Monte-Carlo (MC) simulations show that a major part of particles of a particular physics interest comes from the region of the primary vertex with momentum greater than $0.5\,\mathrm{GeV}/c$. Knowing in advance that a particle is emerging from the primary vertex with large momentum drastically improves the time needed to reconstruct such a track, due to a better track parameter initialization and a small curvature of the track. These observations have led to a decision to split the reconstruction procedure in several stages in order to perform the fast and computationally easy parts first and to reduce the combinatorics in the later stages.

Thus, the CA track finder consists of the following stages (pseudocode line 3):

1. search for tracks of high-momentum ($p > 0.5\,\mathrm{GeV}/c$) quasi-primary (emitted from the target region) particles,

2. search for tracks of low-momentum ($0.1 < p < 0.5\,\mathrm{GeV}/c$) quasi-primary particles,

3. search for tracks of secondary particles with arbitrary momentum,

4. search for tracks with missing hits.

For each set of tracks the same procedure is repeated with a different initial approximation and a different set of cuts. After each stage the input hits included in reconstructed tracks are flagged as used and removed from consideration at the next stage. This approach suppresses possible random combinations of hits and improves the speed of the algorithm at high track multiplicities [3].

After the type of initial triplet parameter initialization (for the primary or secondary track search) and cut values (for the high- or low-momentum track search) are defined by setting the desired track type (pseudocode line 3), the actual triplet building procedure is to begin (pseudocode lines 6–16). Triplets are built station by station, moving upstream starting from the sixth station to the first station.

At the beginning, in order to use Single Instruction Multiple Data (SIMD) registers, all input data is packed into single-precision SIMD vectors, so that all later calculations are vectorized. For reasons of memory optimization, all hits on each station are split into portions of N hits. N is a multiple of the number of elements in the SIMD register, since the hits are processed in a SIMD-ised manner.

Each triplet is built in three steps: to the starting hit second and third hits are added consecutively, rejecting non-physical combinations according to the track model at each step. Thus, at first each hit on a station is considered as a starting hit of a triplet. To the starting hit we add the target with some errors, obtaining in this way a certain spatial direction. This structure is called a singlet. The singlet together with its errors is extrapolated to the next station, taking into account the track model and effects of multiple scattering. While searching for the next hit of this station, the algorithm creates a HitArea object. All hits within the hit area potentially belong to the same track as the singlet hit. Adding each hit from the hit area to the singlet we obtain an array of doublets. The obtained doublets are fitted with the Kalman filter method and some of them are rejected due to a high χ^2-value. In the same manner each doublet is propagated to the next station and the array of triplets for the starting hit is obtained. Most random combinations are rejected due to their χ^2-value while adding the third hit to the triplet.

The triplets discussed above allow to reconstruct tracks with consecutively registered hits in each station. However, because of detector inefficiency some hits may not be registered. In this case a hit in a certain station will be missing. In order to make the algorithm less sensitive to detector inefficiency there is one additional stage of building triplets with one hit missing (Figure 4).

When all triplets are constructed, they have to be be grouped into track candidates. Since momentum is conserved in the magnetic field, in order to find triplets potentially belonging to the same track, one needs to take into account all five state vector parameters. Such so-called neighboring triplets should coincide in position, slope and momentum. The easiest way to require it, is to define triplets as neighbors whenever they share two common hits and have the same momentum within estimated errors (Figure 4). Thus, in the next step the algorithm loops over all triplets and finds and stores for each one the list of possible neighbors.

In order to further simplify the procedure of combining triplets into tracks one can also estimate a possible position of each triplet in the track. This is done with the help of the so-called *level*, which is calculated for each triplet during the search for neighbors. The level of a triplet is defined as the length of the longest continuous chain of neighboring triplets in the direction to the target. The level of a triplet helps to locate the triplet on the track: the greater its value, the longer is the track then can be constructed with the triplet.

Having estimated the position in the track, the triplets can be easily connected to track candidates (pseudocode lines 9–25). The main aim of this procedure is to build the best set of tracks, according to their χ^2-value, that do not share hits. This allows to suppress random combinations of triplets. If two track candidates share a hit, the preference is given to the longest track, since the probability of random hits to be compatible with the track model decreases with the number of hits.

Thus, the procedure starts with building the longest tracks first. Their hits are flagged as used and removed from further consideration. After the longest tracks have been found, the algorithm consecutively searches for tracks one hit less and so on until the number of hits in a primary track is three. Starting with the triplets with the maximum level, a chain of neighboring triplets is constructed, where the level of a triplet is smaller by one than the level of its predecessor. The final chain is considered as a track candidate. A treelike structure of potential track candidates is formed in this manner, from which the best track is selected according to the χ^2 criterion. Having constructed all candidates of a certain length, the algorithm sorts them according to their χ^2. After this procedure is finished, candidates are sequentially checked to contain used hits. If at least one used hit is found, the candidate is discarded; if not, the candidate is stored.

When the stage of track reconstruction for the current set of tracks is finished, the final stage of the algorithm is executed (pseudocode line 27): all the hits used in the reconstructed tracks are removed from the input to the next stage.

2.2 CA track finder algorithm performance

Let us briefly go through the four CA track finder iterations, outlining the initialization parameters and the performance achieved after each iteration.

For evaluation purposes a reconstructed track is assigned to a generated particle if at least 70% of its hits have been caused by this particle. A generated particle is regarded as found if it has been assigned to at least one reconstructed track. If the particle is found more than once, all additionally reconstructed tracks are regarded as clones. A reconstructed track is called a ghost if it is not assigned to any generated particle according to the 70% criterion.

In the first stage the algorithm searches for high momentum primary tracks. Since searching for almost straight tracks originating from the target region is relatively easy due to smaller extrapolation errors and, thus, less combinatorics, this iteration is relatively fast and supposed to reduce combinatorics for the later search. The track finding performance after the first iteration is presented in the table in Figure 5. As one can see, the reconstruction efficiency of this category of tracks 94.9% already after the first iteration, while the reconstruction of low momentum tracks and secondary tracks is not satisfactory. The track reconstruction efficiency as a function of the track momentum shows that the first iteration is able to reconstruct tracks with momenta above 0.5 GeV/c.

The main aim of the second iteration is to search for low momentum primary tracks as well. The resulting performance us shown in the table in Figure 6. After the second iteration, the reconstruction efficiency for the low momentum primary tracks has increased from 56.8% to 91.4%. However, the ghost and clone rate has increased as well because of increased combinatorics. After the second iteration the algorithm is able to reconstruct tracks with momenta down to about 0.1 GeV/c.

The third iteration is targeted at the search for secondary tracks. Figure 7 shows that the reconstruction efficiency of secondary tracks is improved from 84.5% after the second iteration to 86.6% after the third iteration for high momentum tracks, and from 54.2% to 54.7% for low momentum tracks.

In the last iteration the algorithm searches for tracks with missing hits in the STS because of detector inefficiency. The resulting performance is presented in the table in Figure 8. The overall reconstruction efficiency after the last iteration has improved by about 2%.

A special procedure is implemented in the algorithm to suppress clones, merging potentially doubly reconstructed tracks. Also, there is a special extender option, which tries to extend tracks in both directions by searching for unused hits that can be attached to the already reconstructed track. The reconstruction performance after switching on the merger and extender options is presented in the table in Figure 9. The clone rate has decreased by more than a factor of two from 1.0% to 0.4%.

The majority of signal tracks (decay products of D-mesons, charmonia, light vector mesons) are particles with momenta higher than 1 GeV/c, originating from the region very close to the collision point. Their reconstruction efficiency is, therefore, similar to the efficiency of high-momentum primary tracks, which is equal to 97.5%. The high-momentum secondary particles, e.g. from decays of K_s^0 and Λ particles and cascade decays of Ξ and Ω, are created far from the primary vertex, therefore their reconstruction efficiency is lower, namely 91.1%. Significant multiple scattering of low momentum tracks in the material of the detector system and large curvature of their trajectories lead to lower reconstruction efficiencies of 92.6% for primary tracks and of 63.8% for secondary low momentum tracks. The total efficiency for all tracks is 90.9% with a large fraction of low momentum secondary tracks. The rates of clones and ghost tracks are 0.4% and 5.9%, respectively.

2.3 4D CA track finder

Since resolving different physical events in CBM is a non-trivial task, the CA track finder has been further modified to take into account time information. The algorithm was adjusted to take time-slices containing STS hits as input. Each STS hit carries the time measurement t in addition to two spatial coordinates x and y, measured at a certain z-position of the detector plane [4].

Switching to 4D tracking formally implies that the time coordinate be added to the state vector of a track:

$$(x, y, t_x, t_y, q/p) \rightarrow (x, y, t_x, t_y, q/p, t),$$

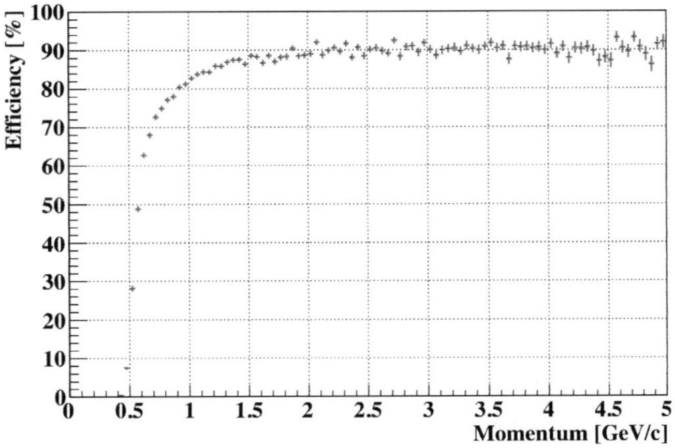

Track category	Eff, %
All tracks	70.4
Primary high-p	94.9
Primary low-p	56.8
Secondary high-p	49.7
Secondary low-p	13.0
Clone level	0.3
Ghost level	0.3
MC tracks found	103
Time, ms/ev	4

Figure 5. Track reconstruction efficiency as a function of track momentum and track finder performance after the search for primary tracks with high momentum.

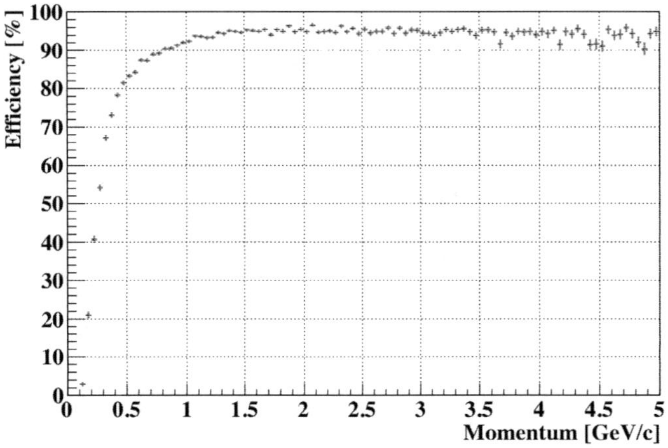

Track category	Eff, %
All tracks	87.8
Primary high-p	95.8
Primary low-p	91.4
Secondary high-p	84.5
Secondary low-p	54.2
Clone level	0.9
Ghost level	5.6
MC tracks found	129
Time, ms/ev	6

Figure 6. Track reconstruction efficiency as a function of track momentum and track finder performance after the search for primary tracks with low momentum.

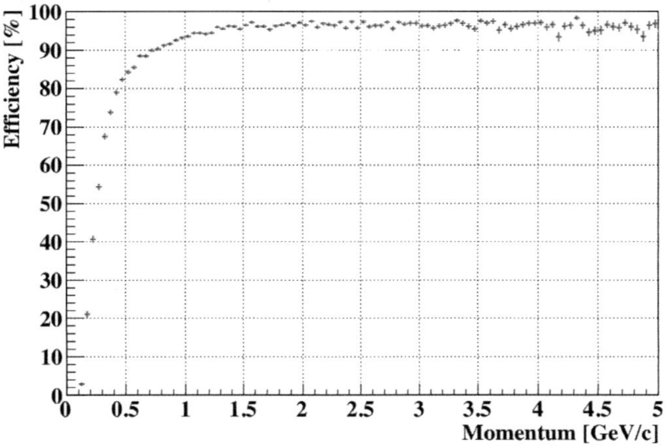

Track category	Eff, %
All tracks	89.2
Primary high-p	97.5
Primary low-p	92.4
Secondary high-p	86.6
Secondary low-p	54.7
Clone level	1.0
Ghost level	5.5
MC tracks found	131
Time, ms/ev	7

Figure 7. Track reconstruction efficiency as a function of track momentum and track finder performance after the search for secondary tracks.

and that all operations with the state vector (e.g. propagation, fitting) involve the time coordinate on the same footing as the spatial coordinates. The feature will be implemented only after taking into account time-of -flight (TOF) detector measurements, since they are several orders of magnitude more precise and should drastically improve the time coordinate resolution. For studies involving the STS detector alone, the time is included into the track parameters, although the propagation and fitting procedures take into account only the spatial coordinates.

The STS hit time measurement information was used in the CA track finder algorithm to improve the speed and the performance. Since the triplets are to be built of three hits, potentially produced by the same particle, these hits should correlate not only in space, but also in time. Neglecting the time of flight between detector planes, hits that belong to the same track should have time measurements that are compatible within the detector time precision. For the sake of speed the additional time dimension was added to the grid structure, so that the time measurement is treated absolutely in the same way

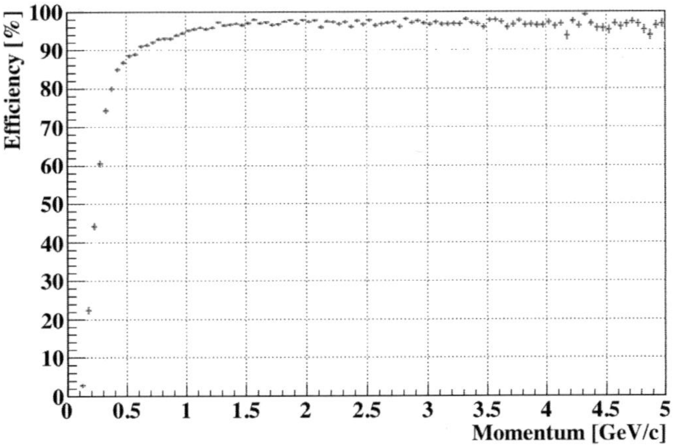

Track category	Eff, %
All tracks	90.9
Primary high-p	97.5
Primary low-p	92.6
Secondary high-p	91.1
Secondary low-p	63.8
Clone level	1.0
Ghost level	5.9
MC tracks found	134
Time, ms/ev	8

Figure 8. Track reconstruction efficiency as a function of track momentum and track finder performance after the search for tracks with missing hits due to detector inefficiency.

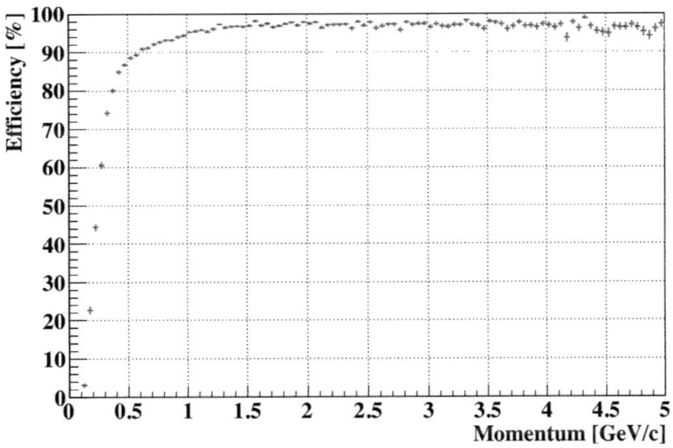

Track category	Eff, %
All tracks	90.9
Primary high-p	97.5
Primary low-p	92.6
Secondary high-p	91.1
Secondary low-p	63.8
Clone level	0.4
Ghost level	5.9
MC tracks found	134
Time, ms/ev	10

Figure 9. Track reconstruction efficiency as a function of track momentum and track finder performance after merging clones.

as the spatial coordinates in the final version of the algorithm. All combinations of hits whose time measurements differ from each other by more than the expected time of flight plus the STS time precision are rejected (Figure 2, cut on the hit time measurement).

After modification of the algorithm the time-based CA track finder was tested to reproduce the results of the event-by-event analysis. Table 1 shows a comparison of the reconstruction efficiencies and the CPU performances between the event-by-event analysis (3D) and time-slice reconstruction with the 4D CA track finder.

As one can see, including the time and optimizing the 3D CA algorithm towards 4D reconstruction have made it possible to achieve a speed comparable to the case of the event-by-event analysis. Moreover, the track reconstruction efficiency has been improved after taking into account the STS time measurement compared with the event-based performance. The efficiency improvement is present even for the extreme case of 10 MHz interaction rate. This can be explained by the presence of slow particles, which create random combinations of hits in case of the event-based approach. These ran-

Table 1. Track reconstruction performance for Au+Au minimum bias events at 25 *A*GeV with the event-by-event analysis (3D) and with the time-slice reconstruction (4D), assuming a 10 MHz interaction rate. Track merging and extending procedures are included.

Efficiency, %	3D	4D
All tracks	90.6	92.2
Primary high-p	97.6	97.9
Primary low-p	93.2	93.5
Secondary high-p	84.4	92.0
Secondary low-p	53.3	65.9
Clone level	8	3.1
Ghost level	7	4.2
MC tracks found	145	145
Time/event/core	11.7 ms	13.6 ms

dom combinations can be rejected in the case of time-slices due to the hit time measurement cut, thus improving the performance.

Conclusion

The First-Level Event Selection (FLES) package for the CBM experiment contains all reconstruction stages: track finding, track fitting, short-lived particles finding, event building and event selection. For the most time-consuming part of the reconstruction procedure, the Cellular Automaton track finder is used. In order to process large time-slices, the CA track finder was vectorized using SIMD instructions and parallelized between CPU cores. The CA algorithm was further adapted to time-slice reconstruction, which is a requirement in case of CBM for the event building. The 4D CA track finder algorithm is now able to achieve the efficiency and timing performance comparable to the event-by-event analysis even at the extreme interaction rate of 10 MHz. The FLES package fulfills the experimental requirements and is ready for time-slice 4D reconstruction in the CBM experiment.

Acknowledgments

The work is supported by BMBF, HIC for FAIR and HGS-HIRe for FAIR.

References

[1] P. Senger, J. Phys. G: Nuclear and Particle Physics **28**, 1869 (2002)
[2] I. Kisel, I. Kulakov, M. Zyzak, IEEE Trans. Nucl. Sci. **60**, 3703 (2013)
[3] I. Kisel, EPJ Web Conf. **108**, 01006 (2016)
[4] V. Akishina, I. Kisel, J. Phys. Conf. Ser. **599**, 012024 (2015), http://stacks.iop.org/1742-6596/599/i=1/a=012024

DOI: 10.1051/epjconf/201612700004

Online and Offline Pattern Recognition in PANDA

Gianluigi **Boca** for the PANDA Collaboration[1,a]

[1] *Dipartimento di Fisica and INFN, Pavia, Italy*

Abstract. PANDA is one of the four experiments that will run at the new facility FAIR that is being built in Darmstadt, Germany. It is a fixed target experiment: a beam of antiprotons collides on a jet proton target (the maximum center of mass energy is 5.46 GeV). The interaction rate at the startup will be 2 MHz with the goal of reaching 20 MHz at full luminosity. The beam of antiprotons will be essentially continuous. PANDA will have NO hardware trigger but only a software trigger, to allow for maximum flexibility in the physics program. All those characteristics are severe challenges for the reconstruction code that 1) must be fast, since it has to be validated up to 20 MHz interaction rate; 2) must be able to reject fake tracks caused by the remnant hits, belonging to previous or later events in some slow detectors, for example the straw tubes in the central region. The Pattern Recognition (PR) of PANDA will have to run both online to achieve a first fast selection, and offline, at lower rate, for a more refined selection. In PANDA the PR code is continuously evolving; this contribution shows the present status. I will give an overview of three examples of PR following different strategies and/or implemented on different hardware (FPGA, GPUs, CPUs) and, when available, I will report the performances.

1 Introduction

PANDA[1] is an experiment that will run at the FAIR accelerator facility in Darmstadt, Germany, starting in 2021. A beam of antiprotons circulating in an accumulation ring with momentum ranging from 1.5 to 15 GeV/c will collide on a hydrogen jet target. The interaction rate at the beginning will be 2 MHz with the goal of reaching 20 MHz when the accelerator delivers the full luminosity.

The PANDA event selection will be completely software. This has been designed to achieve the maximum possible flexibility in the physics channels selection and to exploit the detector possibilities at the fullest. On the other hand this choice poses a great challenge on the Pattern Recognition (PR) software that must keep up with the initial 4 MHz (corresponding to a raw data rate of 20 GB/s) and, in perspective, with the final 20 MHz interaction rate (raw data rate of 200 GB/s), and it must reject the fake tracks produced by the pile-up of hits from previous or subsequent events. A lot of PR code have been developed so far in PANDA. It can be divided essentially in three categories: 1) online PR running on FPGA ; 2) code running on GPU for online/offline PR; 3) code running on CPU for offline PR.

For lack of space only three examples of PR code are described in the following, one running on FPGAs, one for GPUs and one running on CPUs. At the end a short summary of the PR algorithms developed in PANDA can be found.

[a]e-mail: gianluigi.boca@pv.infn.it

First a very brief description of the PANDA central detector is given in the next section in order to allow the reader to understand later the PR algorithms described.

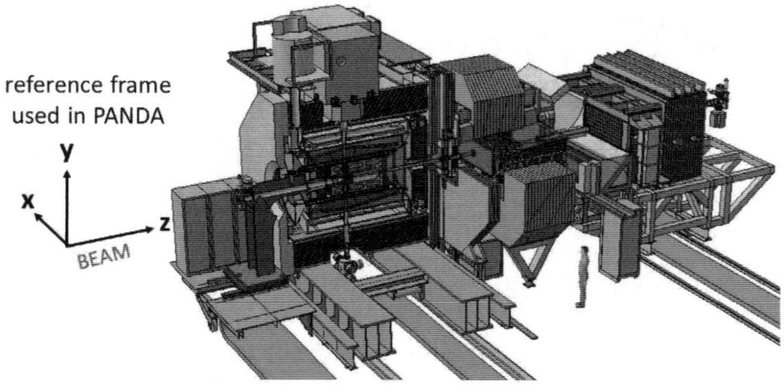

Figure 1. The PANDA full detector

2 The PANDA detector

The PANDA detector is shown in Figure 1. The antiproton beam circulates in the ring along the z-axis in the positive direction. The central part of the detector is inside a 2 Tesla superconducting solenoid. The target system blows frozen hydrogen into the beam line. Inside the solenoid from about 3 cm up to 15 cm away from the beam axis the MicroVertex Detector (MVD) is located, consisting of pixel silicon planes and double-sided strip planes (see Figure 2), the Straw proportional Tubes Tracker (STT) shown in Figure 3, a time of flight tile scintillator detector, a ring imaging silica Čerenkov detector, and a scintillating crystal electromagnetic calorimeter. The muon proportional chambers are inserted between the solenoid iron yoke. The central part is completed by three planes of gas electron multiplier (GEM) detectors covering the very forward angles ($\theta < 11°$), a large backward angle ($\theta > 140°$) scintillating crystal calorimeter and a forward silica disk ring imaging Čerenkov detector. Downstream the solenoid the forward particle section of the experimental setup is placed: a dipole with proportional straw chambers inside, a Time of Flight wall made of plastic scintillators, an aerogel ring imaging Čerenkov, and an electromagnetic and hadronic calorimeter.

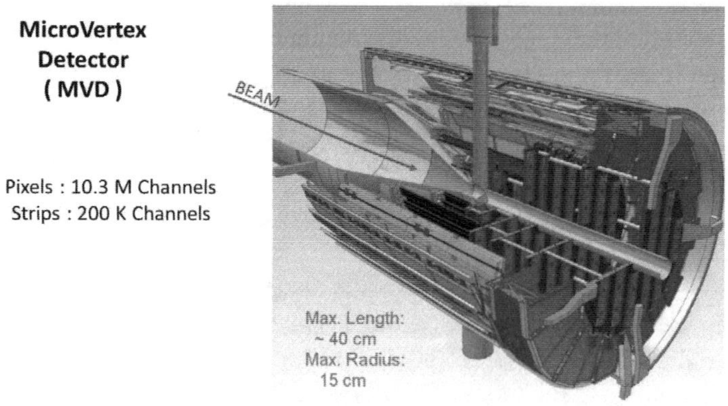

Figure 2. The PANDA MicroVertex Detector

 DOI: 10.1051/epjconf/201612700004

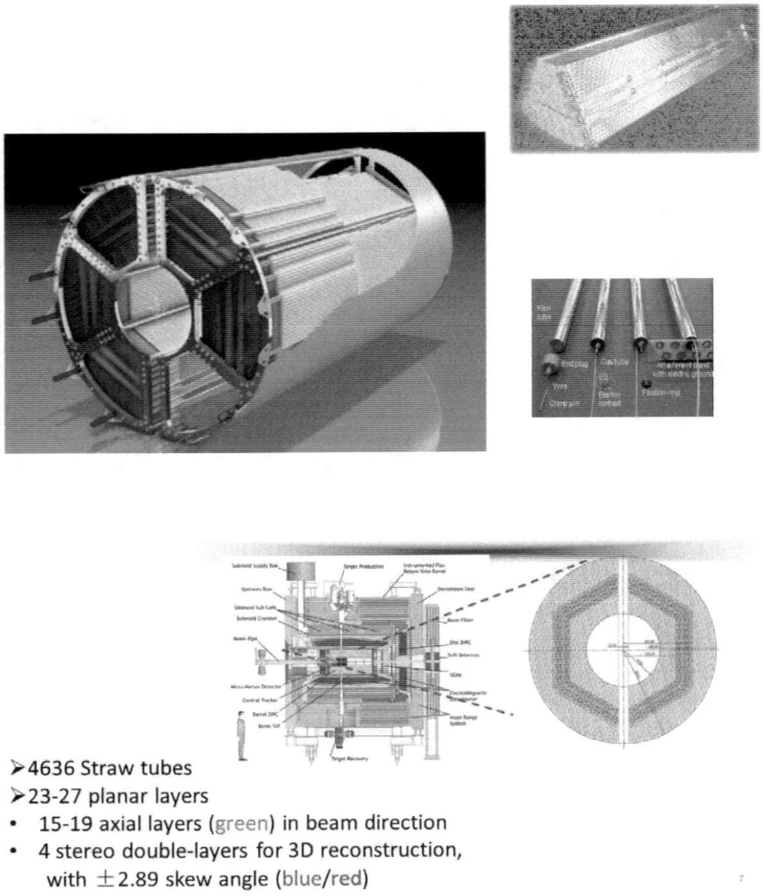

Figure 3. The PANDA Straw Tube Tracker

3 An example of PR running on FPGAs

It is described here an example of a very fast algorithm written for online selection and designed to run on FPGAs. The tracks are searched in the central part of the detector, inside the solenoid, with a road finding algorithm, using only the STT straw hits. First the information on each hit (straw layer and position number, time of arrival of the hit) of the stream of STT hits from the continuous data taking is packed in the FPGA memory in 16 bits per hit (see Figure 4). Then the following algorithms are implemented in the FPGA:

1. The first step consists in finding tracklets with neighbouring STT hits in the x-y-plane combined together (see Figures 3 and 4) starting from the innermost STT axial layers going outwards.

2. The particle trajectory projected on the x-y-plane is a circle; its radius and center position are determined by fitting the tracklets with a circle (χ^2 minimization) in 2 iterations. The $p_\perp$ of the particle at the origin is calculated.

3. The stereo layers hits are associated to the tracklets and the P_z of the particle is found with a χ^2 minimization in two iterations.

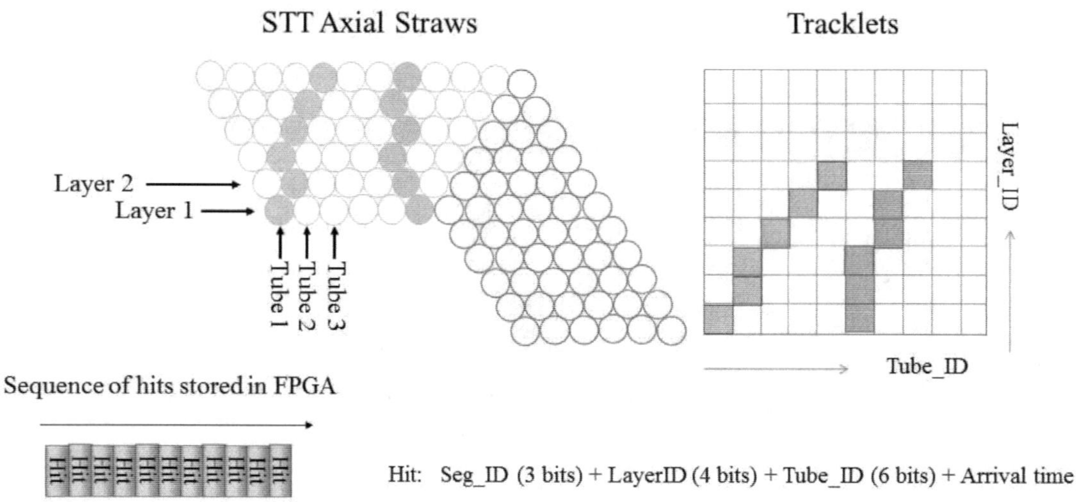

Figure 4. Pattern Recognition code running on FPGAs; on the left top part of the STT axial straws is shown with the convention used to identify uniquely each tube; on the top right an example of tracklets found in the STT detector in the Layer_{ID} versus Tube_{ID} plot. At the bottom the detail of the sequentially stored hit information is shown.

Table 1. FPGA clock counts spent in the algorithm for an event with 6 tracks

	Hit Reading	Road Finder	$p_\perp$ extraction	p_z extraction
Clock Counts	100	$50 \times 6 = 300$	$120 \times 6 = 720$	$120 \times 6 = 720$

The algorithm has been implemented in VHDL language on 140 Virtex5 Xilinx FPGAs and it is parallel. The square root operations and the divisions have been sped up using look-up tables.

The algorithm was tested with a stream of STT hits simulated with the standard PANDA simulation package. Events with six muon tracks were generated uniformly in space. An interaction rate of 20 MHz was assumed, thereby stressing the code at the maximum level possible in the PANDA experiment. The high interaction rate produces also pile-up in the recorded hits as a consequence of the STT maximum drift time ($\approx$ 200 nsec), which is longer than the interaction rate. A detail of the clock counts used in each step is given in Table 1 . The total time per 7-track event is 7 µs ; the momentum resolution obtained is $\sigma_{p_\perp} \approx 3.2\%$ for tracks having $p_\perp = 1$ GeV/c, and $\sigma_{p_z} \approx 3.8\%$ for tracks having $p_\perp = p_z = 1$ GeV/c.

This algorithm will be extended to include the MVD hits in the road finding procedure.

4 An example of PR running on GPUs

I describe here an example of parallel code written for GPUs. The tracks are searched in the central parte of the detector, inside the solenoid, using the STT and MVD hits. The strategy used in this algorithm is based on a Hough transform method. After the information of the STT and MVD hits are loaded in the GPU memory the algorithm executes the following steps:

1. in the x-y-plane the x and y coordinates of all MVD hits are transformed to coordinates u and v by a conformal transformation:
$$u \equiv \frac{x^2}{x^2+y^2}; \qquad v \equiv \frac{y^2}{x^2+y^2} \tag{1}$$
consequently the circular trajectory originating from (0,0) in x-y-plane transforms into a straight line going through the MVD points.

2. The same conformal transform is applied to the drift circles of the STT Axial straw hits. Those drift circles transform again to circles in the u-v-plane that are tangent to the particle trajectory (see Figure 5).

3. The Hough plot in the variables θ and R is constructed using the MVD hits and the STT axial straw hits. For each MVD hit a bundle of straight lines (in the u-v-plane) passing through the MVD hit is generated, with the following parametrization:
$$u_i \cos\theta + v_i \sin\theta = R,$$
where u_i and v_i are the MVD hit coordinates. For each STT axial straw hits a similar parametrization is used:
$$u_i \cos\theta + v_i \sin\theta = R \pm d,$$
where u_i and v_i are the coordinates of the center of the straw and d the drift radius of the hit.

An example of the resulting Hough plot for a single-track event is shown in Figure 6. The accumulation point is highlighted in the figure. It is obtained by a peak finding algorithm and allows the determination of the radius and center of the trajectory helix. In the figure the Hough plot consists in 1800×1800 cells in the R, θ variables. This algorithm was parallelized and run on NVIDIA Tesla

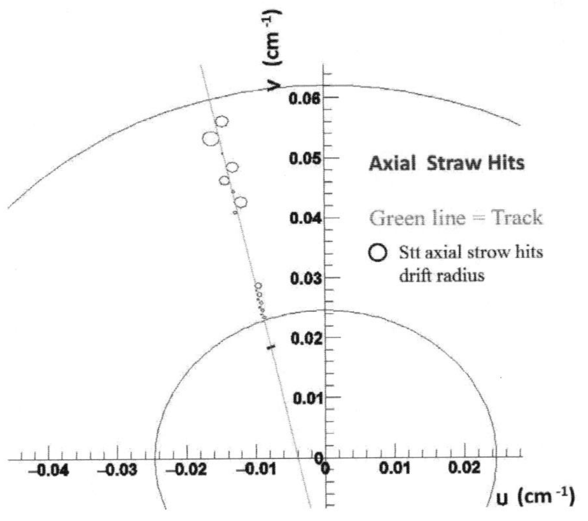

Figure 5. Simulation of a particle trajectory produced at the origin in x-y-plane and conformal-transformed as described in the text

DOI: 10.1051/epjconf/201612700004

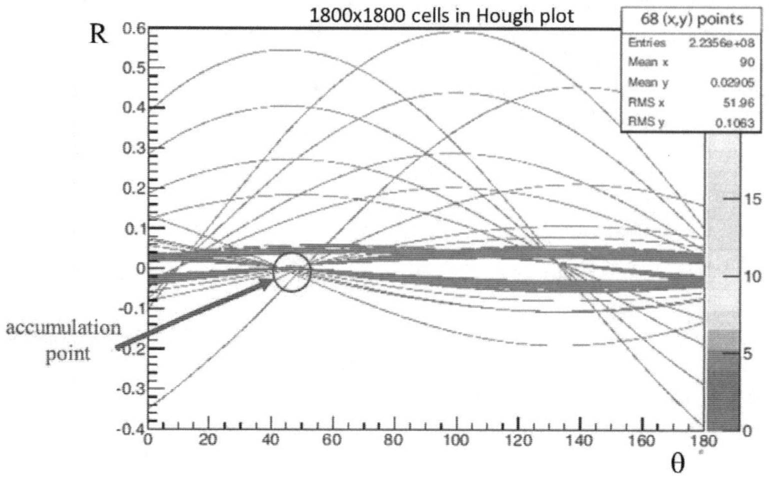

Figure 6. Hough plot obtained with MVD + STT axial straws hits in a 1800 × 1800 Hough plot.

K20K hardware driven by a dedicated CPU. It was written in plain CUDA. The data was fed in the CPU first and then into the memory of the GPU. The performances are relative to the PR algorithm run with the data deposited in the shared memory of the GPUs and are shown in Table 2. So far this algorithm delivers only the $p_{\perp}$ of the tracks; it will be completed with the inclusion of the calculation of p_z of the tracks, using MVD hits and STT stereo hits.

Table 2. Performance of the parallel algorithm run on NVIDIA Tesla K20K GPUs.

	Hough plot size (N×N) 256 cells	Hough plot size (N×N) 512 cells	Hough plot size (N×N) 1024 cells	Hough plot size (N×N) 2048 cells	Hough plot size (N×N) 4096 cells
% reconstructed tracks	33.0	41.0	55.4	76.0	84.4
% failed tracks	67.0	59.0	44.6	24.0	15.6
% spurious tracks	27.4	14.8	7.2	3.5	0.6
time per event (ms)	0.006	0.013	0.030	0.117	0.507

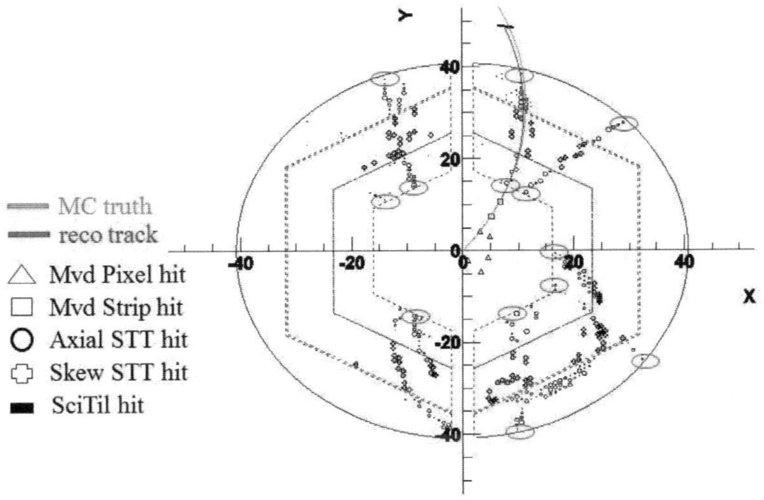

Figure 7. An example of a PANDA single track event projected on the x-y-plane. The green line is the true simulated track, the red line is the reconstructed tracks. Most of the hits in this figure are spurious, remnants of previous (or subsequent) events. Here the (extreme) case of an interaction rate of 20 MHz is show, while at the beginning the PANDA interaction rate is only 2 MHz. The red-circled hits are the STT axial hits from which the tracklets are started.

5 An example of PR running on CPUs

I describe in this section a PR algorithm using the central detector hits to find tracks. It combines the road-finding strategy with the Hough method to maximize efficiency and rejection of spurious hits. Initially it was designed for offline PR and run on CPUs. However, parallelization and implementation on GPUs is foreseen for a possible use online as well.

The first step is finding the tracklets in x-y-plane by using the STT axial straw hits only. They are formed starting from hit at the boundary of the STT detector (red-circled hits in Figure 7) and looking for adjacent hits. In the figure one can notice the large number of spurious hits not belonging to any real track. Those are caused by remnants of tracks from previous events of subsequent events. In PANDA it is one of the major problems encountered in a PR algorithm using the STT detector. It is caused by its large maximum drift time (about 200 ns). Figure 7 is an example of pile-up event in the (extreme) case of interaction rate of 20 MHz, a situation possibly occurring only at a later stage of PANDA. At the start-up of PANDA the interaction rate will be only 2 MHz.

The algorithm proceeds to calculate the conformal transformation in Equation (1), thereby transforming the circular trajectory in the x-y-plane into a straight line in the u-v-plane. A first fast calculation of the radius and center of the trajectory helix is done with a Hough transform plot on the straight trajectory in the u-v-plane. Using this first result the algorithm attempts to add to the tracklet more STT axial straw hits and MVD hits close enough to the trajectory. A χ^2-minimization fit to all STT+MVD hits is performed and the $p_\perp$ and charge of the particle are finally established.

The p_z is determined by associating the STT stereo straw hits to the helix. The stereo angle is only 3° and helps selecting the true hits belonging to the tracks. In fact the spurious straws intersect the trajectory cylinder in unphysical regions (too far back or too far forward). When projected on the trajectory cylinder the trajectory is a straight line (see Figure 8). The MVD hits look like points,

DOI: 10.1051/epjconf/201612700004

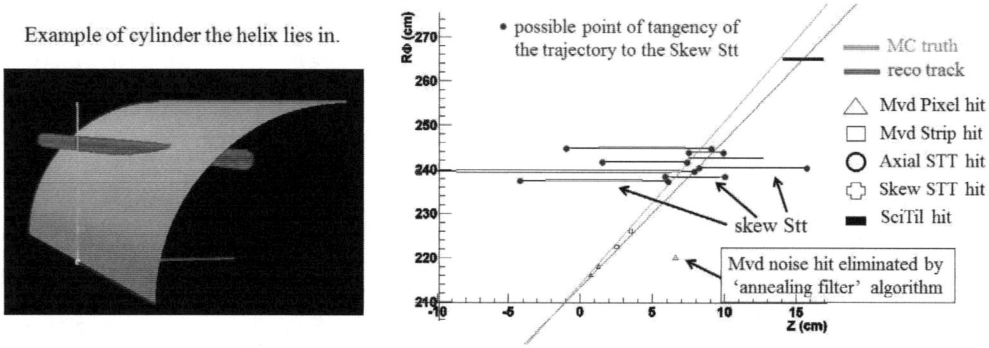

Figure 8. Left: cartoon showing an example of the intersection of a stereo STT straw tube (red cylinder) with the cylinder on which the helix trajectory lies (yellow surface); right: projection on the later surface of the helix cylinder of a helix trajectory. The stereo STT straw (≡ skew STT) intersections look like ellipses squeezed in the vertical direction because of the small stereo angle (3 degrees). One can notice one MVD spurious hit that was eliminated by an annealing filter type of algorithm.

while the stereo STT straw tube intersections look like ellipses squeezed in the vertical direction because of the small stereo angle (3 degrees), with two possible points of intersection with the helix trajectory (blue points in Figure 8). A straight line χ^2-fit delivers the p_z of the track. All possible combinations of stereo STT straw tube intersection points are taken into account in the fit and the solution corresponding to the best χ^2 is chosen. At this point the trajectory is completely determined and a last round of elimination of spurious hits is performed with a simple proximity selection.

Because of the pile-up mentioned earlier, the algorithm may find many spurious tracks (Figure 7 in the worst case of 20 MHz interaction rate). Consequently two cleanup procedures have been written rejecting a track if its hits are not continuous (spatially) in the MVD detector (in the first procedure) or in the MVD and STT detectors (in the second procedure). The performance of this PR is summarized in Tables 3 and 4, calculated with simulated events in which an interaction rate (and consequent pile-up of hits) of 2 MHz was assumed. Events were simulated with 1 to 8 (muon) tracks. All results are summarized here for the two cleanup procedures.

The CPU time spent per track per event on an Intel i7-2600K 64 bits 3.4 GHz core is shown in Table 5 for 2 MHz and 20 MHz interaction rate. The code is not yet parallelized and another factor of 5–10 in speed is expected to be gained when running on a GPU.

6 Other algorithms implemented in PANDA

In PANDA many other Pattern Recognition or tracking algorithms have been written during the years and unfortunately describing them would require too much space. The following is a partial list of them.

- A PR using the MVD hits to find radius and center of the helix cylinder of a track not necessarily produced at the vertex. This algorithm is based on the projection of the circular trajectory on a Riemann sphere (or paraboloid) and can in principle be used to find also secondary vertexes.
- Cellular Automaton types of algorithms that use the MVD and STT hits have been written and implemented on CPUs and GPUs;
- A Hough transform method implemented on GPUs that uses MVD and STT axial straw hits only. This procedure uses an alternative parametrization of the particle circular trajectory in the x-y-plane.

Table 3. Performance of the algorithm run on CPUs assuming an interaction rate of 2 MHz.

MVD hit only cleanup procedure	
True Tracks found	from 92.6% to 98.9 %
STT True Hits found in True Tracks	from 97.8% to 99.7 %
MVD True Hits found in True Tracks	from 98.2% to 99.7 %
Spurious Tracks per event	from 0.26 to 0.33
CPU time per event per track	from 0.26 to 0.33
MVD+STT hit cleanup procedure	
True Tracks found	from 92.2% to 98.5 %
STT True Hits found in True Tracks	from 98.7% to 99.7 %
MVD True Hits found in True Tracks	from 98.2% to 99.7 %
Spurious Tracks per event	from 0.07 to 0.11
CPU time per event per track	from 0.26 to 0.33

Table 4. Performance of the algorithm run on CPUs assuming an interaction rate of 20 MHz.

MVD hit only cleanup procedure	
True Tracks found	from 92.2% to 98.5 %
STT True Hits found in True Tracks	from 95.6% to 98.5 %
MVD True Hits found in True Tracks	from 94.8% to 98.6 %
Spurious Tracks per event	from 2.65 to 2.77
MVD+STT hit cleanup procedure	
True Tracks found	from 66.7% to 77.5 %
STT True Hits found in True Tracks	from 95.7% to 97.2 %
MVD True Hits found in True Tracks	from 95.1% to 97.6 %
Spurious Tracks per event	from 0.58 to 0.89

- An algorithm using only STT hits to find the T_0 ($\equiv$ the time when an event was produced in the primary interaction) of an event; it was implemented on CPUs and also on GPUs.
- Algorithms written in order to find the secondary vertexes;
- PR code for tracks hitting the GEMs (inside the solenoid, in the forward direction).
- PR code for the very forward part of the PANDA detector (trackers in the dipole magnet).

DOI: 10.1051/epjconf/201612700004

Table 5. CPU time consumption per track per event for an interaction rate of 2 and 20 MHz; MVD+STT hit cleanup procedure.

	2 MHz interaction rate	20 MHz interaction rate
1-track event	1.6 msec	6.5 msec
4-track event	5.0 msec	9.5 msec
8-track event	9.5 msec	13.5 msec

References

[1] PANDA Collaboration (M.F.M. Lutz et al., *Physics Performance Report for PANDA: Strong Interaction Studies with Antiprotons* (2009), arXiv:0903.3905v1 [hep-ex]

 DOI: 10.1051/epjconf/201612700005

First Results of an "Artificial Retina" Processor Prototype

Riccardo **Cenci**[1,3,a], Franco **Bedeschi**[3], Pietro **Marino**[1,3], Michael J. **Morello**[1,3], Daniele **Ninci**[1,2], Alessio **Piucci**[1,2], Giovanni **Punzi**[1,2], Luciano **Ristori**[4], Franco **Spinella**[1], Simone **Stracka**[1,2], Diego **Tonelli**[5], and John **Walsh**[1]

[1] *INFN Sezione di Pisa, Italy*
[2] *Università di Pisa, Italy*
[3] *Scuola Normale Superiore, Pisa, Italy*
[4] *Fermi National Accelerator Laboratory, Batavia, IL, USA*
[5] *CERN, Geneva, Switzerland*

Abstract. We report on the performance of a specialized processor capable of reconstructing charged particle tracks in a realistic LHC silicon tracker detector, at the same speed of the readout and with sub-microsecond latency. The processor is based on an innovative pattern-recognition algorithm, called "artificial retina algorithm", inspired from the vision system of mammals. A prototype of the processor has been designed, simulated, and implemented on Tel62 boards equipped with high-bandwidth Altera Stratix III FPGA devices. The prototype is the first step towards a real-time track reconstruction device aimed at processing complex events of high-luminosity LHC experiments at 40 MHz crossing rate.

1 Introduction

With the luminosity increase expected at the Large Hadron Collider (LHC) in the next years, the available amount of collected data will scale by more than one order of magnitude. For many years, advancements in electronics technology (i.e. Moore's law) were able to cover for most of bandwidth increase due to higher luminosity and higher number of channels. However, this is not valid anymore, and today's technology advancements, when available, carry an additional development cost (compare a simple clock speed increase to software development for multi-core CPU), while physics complexity is still growing (luminosity, event pile-up, higher precision measurements). In some cases big steps forward in algorithms and architectures are already required, but it will be the default when moving to LHC Phase-2.

While in the past for one-collision events requirements on simple quantities (e.g. total calorimeter energy, number of muons, high transverse-momentum tracks) were sufficient to lower the bandwidth, multiple-collision events can contain interesting processes in almost any event, and different solutions, like real-time reconstruction, need to be explored. Real-time reconstruction of charged particle trajectories can help selecting events and reducing data size, but it will bring larger combinatorial problems, that require higher parallelization to be solved within typical latencies. In the '90s the Collider Detector at Fermilab (CDF) demonstrated the possibility to reconstruct in real-time two-dimensional tracks

[a]e-mail: riccardo.cenci@pi.infn.it

from clusters of aligned hits in a large drift-chamber [1]. The algorithm used by CDF was based on matching hits in the detector with precalculated patterns, and an updated version of the same approach is at the root of other current projects for LHC experiments [2, 3]. Performing a similar task at the LHC crossing rate is very problematic, due to the large combinatorial and size of the associated data flow that would require a massive network infrastructure and computing power. The ATLAS experiment is implementing an updated version of the CDF algorithm [2], called Fast TracKer (FTK). The typical number of clock cycles required by the FTK trigger to reconstruct an event is around 20,000, while the same number for a low-level trigger at the high-luminosity LHC is estimated to be around 25, so this solution cannot be used at the first-level trigger. An alternative approach developed for the CMS experiment is focused on reducing combinatorial locally at the detector level using pairs of tracking layers at short distance (1-4 mm) [4]. The local readout electronic searches for pairs of hits (called stubs), and sends out only the ones corresponding to tracks with transverse momentum greater than 2 GeV/c.

A new pattern-matching methodology has been recently proposed under the name "artificial retina" algorithm [5], inspired by the quick detection of edges in the visual cortex of mammals. Its aim is to further increase the parallelism of the pattern-matching process and decrease the number of stored patterns, to reduce latencies and hardware size. The purpose of the INFN-Retina project described below is to explore the possibility of using the artificial retina algorithm for a detailed tracking reconstruction at 40 MHz, compatible with the first step of online event selection at LHC, at a reasonable cost. A hardware prototype based on already-existing FPGA readout boards is under development, to explore the potential of this new approach in a realistic experimental environment, albeit at lower rates. The main purpose of the prototype is to show the possibility of reconstructing quality tracks using the same hardware resources that are normally required just for reading out the raw hits.

2 The "artificial retina" algorithm

The "artificial retina" algorithm, elsewhere referred to as retina algorithm, was inspired by the first stage of mammal vision. Recent experimental studies show that specific neurons, called receptive fields, receive signals only from specific regions of the retina, in order to reduce the connectivity and save bandwidth. The neurons are tuned to recognize a specific shape and the response is proportional to how close are the stimulus shape and the shape for which the neuron is tuned to. Generated in parallel, the responses of neurons are then interpolated to create a preview of image edges in about 25 ms, corresponding to about 25 clock cycles and perfectly matching the requirement for processing an event mentioned above. Therefore those concepts of early vision can be used to realize a viable highly-parallel implementation of an "analog" pattern-matching system, where each pattern is assigned a continuous level of "matching", rather than a simple binary response. The mathematical aspects of the algorithm have some similarities with the "Hough transform" [6, 7], a method already applied for finding lines in image processing; however, the main challenge here is the design of the physical layout and the development of an implementation capable to sustain the event rate at high-luminosity LHC experiments [8].

For configuring the algorithm, the space of track parameters is divided into cells, which mimic the neurons connected to the receptive fields of the retina. The center of each cell corresponds to a specific track in the detector that intersects the layers in spatial points called *receptors*. A first mapping connects each cell with the receptors, as shown in Figure 1. For a group of contiguous cells, where variations of track parameters are small, the corresponding receptors in the detector layers would belong to a limited area. A second mapping, also shown in Figure 1, connects clusters of cells to

DOI: 10.1051/epjconf/201612700005

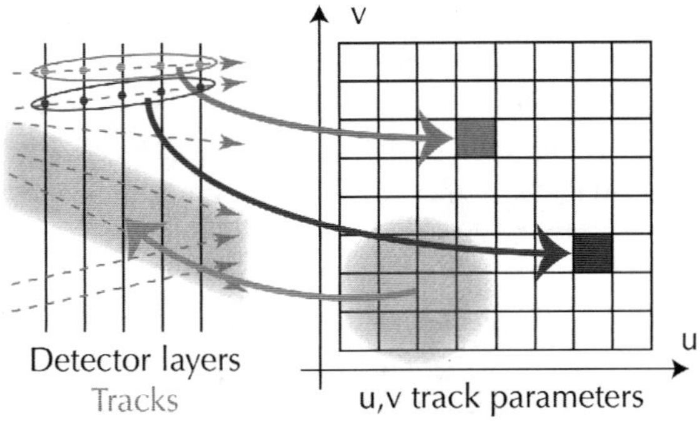

Figure 1: Retina mappings for tracks on a plane without magnetic field, where tracks can be described by two parameters u and v.

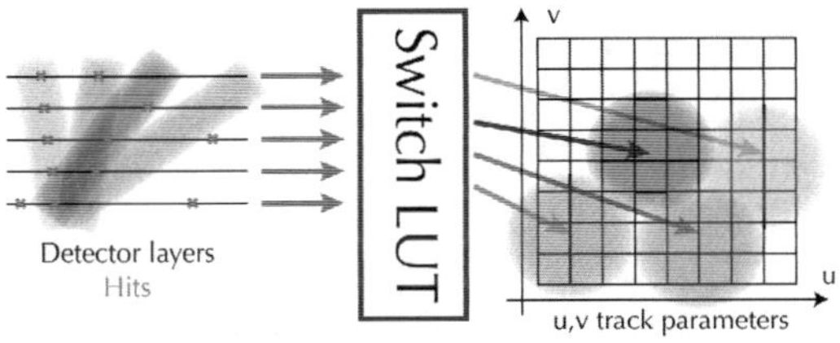

(a) Step 1

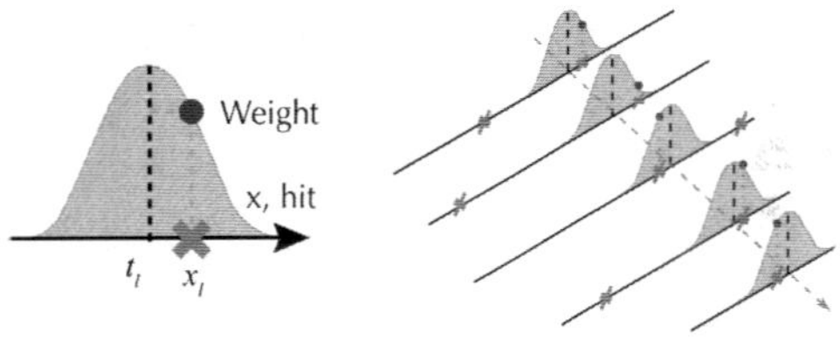

(b) Step 2

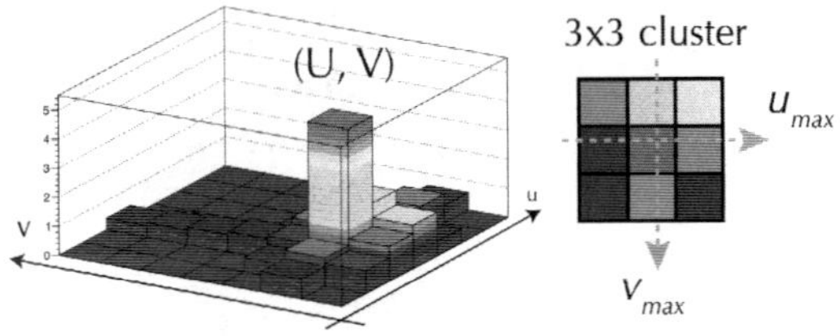

(c) Step 3

Figure 2: Processing steps of track reconstruction with the artificial retina algorithm.

areas of the detector, and this information is recorded in a look-up-table (LUT). A commercial PC is enough to generate the two mappings starting from simulated tracks.

The track reconstruction, to be implemented on high-speed FPGA devices, has three steps. During step 1, also referred to as the *switching* step, detector hits are distributed only to a reduced number of cells, according the LUT's created in the configuration phase, as shown in Figure 2a. In step 2, sketched in Figure 2b, for every incoming hit a Gaussian weight w is accumulated in each cell by a logic unit called "engine". The weight is proportional to the distance to the receptor and defines as follows:

$$w = \exp\left(-\frac{d_l^2}{2\sigma}\right)$$

where d_l is the distance, on the layer l, between the hit and the corresponding receptor, and σ is a parameter of the algorithm, that can be adjusted to optimize the sharpness of the response of the receptors. After all the hits from the same event have been processed, in step 3 tracks can be identified looking for local maxima of the accumulated weight over the cell's grid. For a track resolution similar to offline reconstruction, the grid does not require a high granularity, because significantly better precision can be obtained by computing the centroid of the accumulated weights for the cells surrounding each maximum, as shown in Figure 2c.

Compared to other algorithms, the retina method takes advantage of two levels of parallelization. First, each cell processes in parallel hits from a limited region of the detector, reducing the required input bandwidth for the single cell. Moreover, if time information is associated to every hit and weights are accumulated separately for a small number of consecutive events, events can be processed simultaneously. The latter allows to feed data continuously to each cell, because cells are not receiving the same number of hits for every event.

Another important feature is the total system bandwidth: the bandwidth increases significantly during the hit distribution, because multiple copies of the same hit can be produced, but shrinks down when only the information about local maxima is kept in the last step. The larger bandwidth can be physically managed only because it is limited to one stage of the system. Curiously, a similar behavior of the bandwidth is found in the first stage of mammal vision as well.

3 Architecture of the track processing unit

The implementation of the retina algorithm on a real device is called track processing unit (TPU). The first step of the algorithm is achieved using a switching network able to redirect hits from the readout module to any engine of the system. The network increases the global bandwidth making multiple copies of hits, but it never sends a hit to all the engines, reducing the input bandwidth for them. The network is assembled using basic 2-way dispatchers (2d) with two inputs and two outputs: data on any input can be redirected to any output using two splitters and two mergers, as shown in Figure 3a. To implement a dispatcher with 2^n inputs/outputs, shown in Figs. 3b and 3c, we need N 2-way dispatchers connected together, where

$$N(n) = 2N(n-1) + 2^{n-1}.$$

The dispatchers work as a pipeline and the latency is proportional to n. The space parameters grid is divided into matrices of engines that fit inside a single device. Maxima can be found on any cell of the grid keeping a one-cell overlap between the matrices. The switching network can be easily scaled adding enough inputs to receive data from the readout modules and outputs to transfer data to all the devices used for the engines.

The best device to implement a prototype of the TPU system is FPGA, due to their high flexibility, and shorter time for designing and test compared to standard ASIC devices. Moreover, FPGA performances are still increasing at a steady pace, taking advantage of new silicon technology (14-16 nm), increasing the number of logic elements (up to 5.5 millions), and including faster connection technology (up to 30 Gbit/s on a single serial link). After testing the system performances on FPGA, the firmware can be always transferred on ASIC devices for mass production. Preliminary studies shows that a TPU system for a modern detector can be implemented using 128 FPGA devices [9], and detailed numbers for the estimated resources and cost are reported on Table 1. All the required devices are already available on the market, and devices available in few years can improve the performances by a factor ten.

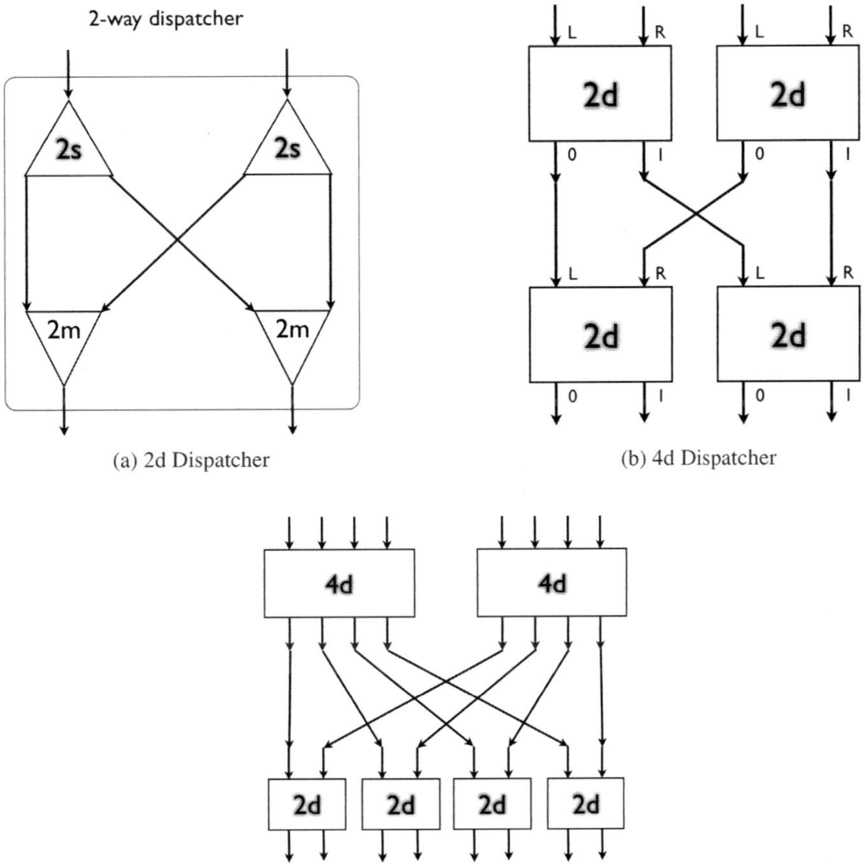

(a) 2d Dispatcher (b) 4d Dispatcher

(c) 8d Dispatcher

Figure 3: Schemes of dispatchers with 2, 4, and 8 inputs/outputs.

4 First prototype design

4.1 Geometry details and simulation studies

To validate the algorithm and measure its performances, a TPU prototype was designed to reconstruct tracks in a small tracker. A telescope with six single-coordinate layers (50 cm long, 20 cm wide) and no magnetic field [10], shown in Figure 4, was used as model. Any track going through the telescope has been parametrized with two variables: the x coordinates on the first and last layer, below referred to as u and v. The GEANT4 simulation of the tracker [10] shows that real tracks are likely to be along z-axis, corresponding to the region around the diagonal in the u-v-plane.

Fundamental design parameters of the system are: the minimum number of cells needed to cover the space parameters and to obtain track resolution; the size and inter-connectivity of the system. The retina algorithm for the telescope was simulated using a C++ emulator [9], showing that resolution and efficiency similar to offline tracking systems can be achieved using only 3,000 cells to cover the diagonal region mentioned above. The emulator was also used to compute the switching network configuration, and as debugging tool, providing data at different stages of processing. The prototype

Table 1: Estimation of resources and cost for a track processing unit that can reconstruct tracks in a modern HEP detector at LHC.

crossing frequency	40 MHz	HL-LHC design
# of layers	10	2 in fringe B field
number of hits per layer per crossing	300	Lumi = 10^{33}-$10^{34} cm^{-2} s^{-1}$
number of bits per hit	40 bits	2D pixels + time
total hit bandwidth	5 Tbit/s	(readout)
number of engines	50k	1/20 of precalculated patterns used by CDF
engines per FPGA	800	
number of FPGAs (1M logic elements)	64+64	switch + engine boards
links per FPGA	64	@10 Gbit/s (standard)
max total bandwidth	40 Tbit/s	max 8 copies per hit
clock speed	500 MHz	
cycles to process one event	150+350	switching + engine steps
latency	1 µs	
number of events in pipeline	16	
cost	1 M€	5 k€ switch FPGA, 10 k€ engine FPGA

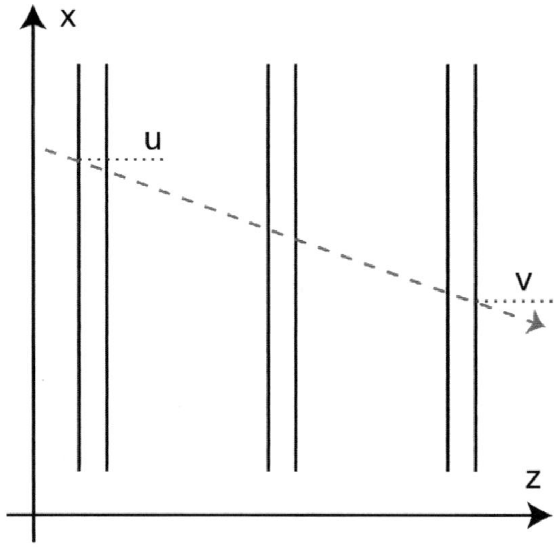

Figure 4: Sketch of the 6-layer telescope used for the prototype, showing also the two parameters u and v used to represent a track.

was designed to be implemented on a currently-available electronic board (see description below) that includes several large and high-bandwidth FPGA devices. The diagonal region in the u-v-plane was covered using matrices of cells, each one of them fitting inside a single device of the chosen board. The matrices in different devices overlap by one cell to be able to find local maxima in all the cells.

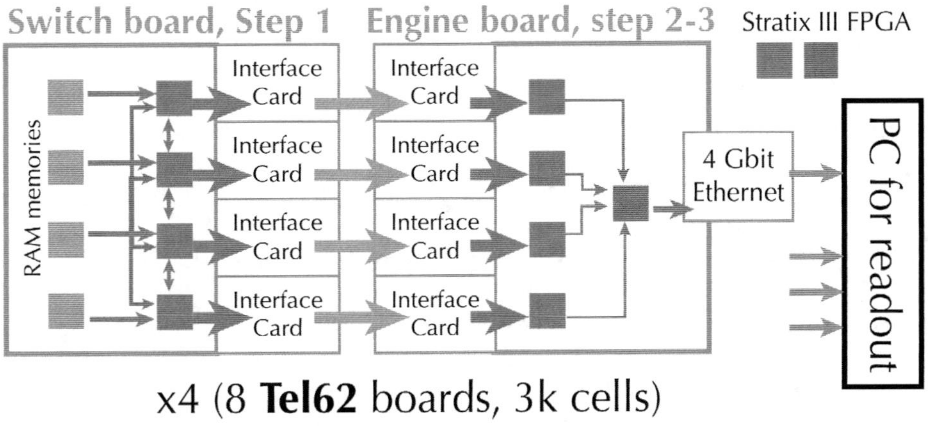

Figure 5: Architecture of the first prototype using Tel62 boards.

4.2 Board details

The board Tel62, developed by INFN Pisa for the NA62 experiment [11], was chosen to implement the TPU prototype because it carries multiple large and high-bandwidth FPGA chips and provides fast interconnection between them. The board includes four Stratix III chip for data processing, each with approximately 200k logic elements. Each chip is connected through a high-speed link (10 Gbit/s) to a master chip (another Stratix III) that collects the processed data and controls the other FPGAs. The main clock is 40 MHz, while the processing internal to the FPGAs is done at 160 MHz. Each processing chip is connected to a 2 Gbyte DDR2 RAM through a high-speed bus (40 Gbit/s), to a mezzanine connector that can host various interface cards for I/O (5 Gbit/s), and to the neighbouring FPGAs (2.5 Gbit/s). The master FPGA is also connected to a mezzanine card with four Ethernet ports (2.5 Gbit/s). Implemented on an embedded PC that sits on the board, a slow-control interface can be used to monitor all the devices and to perform various standalone tests.

4.3 Algorithm implementation

The Tel62 board is designed for a standard DAQ system, where bandwidth typically reduces following the data-flow (only selected data move to the next step) and data streams do not have many connections between all of them. In any TPU system, however, the bandwidth increases in the first step and data have to be exchanged also laterally within the switching network. Therefore, two boards were connected together: the data-flow was reversed in the first one, where the switching network was implemented (switch board), and a newly designed interface card was used to connect a second board, where the processing engines were implemented (Engine board), as shown in Figure 5. This configuration is fully consistent with the implementation of the system proposed in Ref. [9], so the prototype results are meaningful for future applications to real detectors.

The processing engine is implemented in a fully pipelined mode, so each engine can receive one hit for each clock cycle [12]. All the engines in a chip receive the same hit sequence and, after a special end-event (EE) hit is received, engine values are copied to the following stage where the search for local maxima starts, while the next hit sequence enters. The latency between the input of EE hit and the output of maximum data is fixed. As explained in more detail below, we fit approximately 200 engines on each Stratix III device (EP3SL200F1152), therefore 16 chips, corresponding to four pairs of Tel62 boards, are enough to implement the whole system.

4.4 Hardware development details

The firmware for the different FPGA chips on the boards was developed using the software "HDL Designer" by Mentor Graphics and the Altera specific software "Quartus II". Most of the firmware is written in VHDL using generic components configurable through parameters, so they can easily be reused for systems on larger devices. The same firmware is loaded into all the four chips on the boards, so specific values for engine receptors and switching network LUT's have to be loaded through the slow control and stored on internal RAM's.

The firmware for the switch board chips is based on a 4d dispatcher block with four inputs and four outputs already described above. Each splitter in the basic 2d dispatcher copies the input data to zero, one, or both outputs, according to a 2-bit LUT value based on the hit coordinate and layer. Additional logic is also required to implement the correct propagation of EE hits: generated by the readout, they arrive separately on each input, and, after the switching step, hit sequences on each output have to contain all the appropriate hits from the event, tailed by a single EE hit. Each of the four switching block outputs corresponds to a different chip on the engine board. Because each engine chip is connected only to one switch chip, all the switch outputs between chips are properly redirected, using the lateral connections or through the master chip. Then, the data streams are merged from the redirected switch outputs and transferred to the engine board through the interface card, as shown in Figure 5.

The firmware for the processing chips of the engine boards is based on a matrix of engines. Each engine completely surrounded by other ones has additional logic to check if the accumulated value is greater than the first neighbors one. The maximum search requires a large number of interconnections between engines, limiting the number of engines that can be fit in our Stratix III FPGA to a 16x15 matrix[1]. A priority encoder manages the serial transfer of resulting maxima to a single output FIFO. Due to the reduced bandwidth achieved at this stage, maxima from all the processing chips are moved to the master chip, and then sent out to a PC through the Ethernet connectors.

5 First results

The prototype is an advanced stage of development and some tests were already performed on a switch board connected to an engine board. The basic firmware for all the system devices has been written and the logic behavior verified using the "ModelSim" software by Mentor Graphics. Simple hit sequences were loaded on internal FIFO's of the switch board chips, and the hits were correctly dispatched to the proper output channels, together with the EE flags. During these tests, all the logic for the switching, internal and between chips, was running properly at the maximum clock rate of the board (160 MHz). The data transmission between the switch and engine boards was successfully run at 80 MHz, using two data channels for each interface card pair. Hit sequences were also loaded on the input FIFO of the engine chip: the hits were processed and the correct maxima data were received by the readout PC through Ethernet connection. Also here the internal logic for the engine is running properly at 160 MHz clock rate. Because the engine input latency is proportional to the number of hits in the sequence and having hit sequences for realistic events with an average of 70 hits per layer, the maximum rate sustainable by the engine board is 1.8 MHz with a latency smaller than 1 μs. The final integration and tests of the firmware are currently ongoing.

[1]The place and route tools from Quartus II report 90% of resources used on our specific device.

6 Future plans

After validating the algorithm and measuring its performances for a small tracker with available boards, another prototype will be built in the next months to test the TPU performances with data from a 3D tracker partially inserted in a magnetic field. For this geometry the algorithm requires three additional parameters to describe a track, but considering a parameter space with five dimensions will increase too much the number of cells. If variations for some parameters can be approximated to small perturbations, we can factorize the parameter space into the product of two subspaces [9]. Therefore we perform the full pattern recognition using only a two-dimensional grid for two parameters, and then extract the other ones using lateral cells (six of them for each cell of the main grid) that represent a track with the same values for the first two parameters and slightly different ones for the remaining three. The lateral cells and the computation of 3D distances require additional logic, so the engine implementation has to be extended and the maximum number of engines fitting in one chip has to be updated. The fundamental parameters of the prototype are the minimum clock speed to reconstruct tracks at LHC Run3 rate, and the size and throughput of the FPGA device needed to implement the system. The prototype will consist only of a basic unit of the system implemented on an ASIC prototyping board with two very large and very high-bandwidth FPGA's. Multiple connections between the two devices will allow to test different configurations, e.g. a standalone switching network connected to a standalone engine matrix, both receiving data at full speed prepared using the algorithm emulator.

7 Conclusions

Computing and storage available for future experiments at high-luminosity LHC will not be able to cope with the increase of data rate, therefore more processing will have to be performed "online" to reduce event rate and size. The methods already employed for data triggering may not scale well and alternative advanced solutions should be explored, like the "artificial retina" algorithm, that exploits higher degrees of parallelization and provides analog response. The INFN-Retina project aims to demonstrate the feasibility at reasonable cost of a system based on this algorithm able to reconstruct tracks at rates expected for LHC Run3. A first, sizable hardware prototype of a retina tracking system with 3,000 patterns is under advanced development, based on already existing FPGA readout boards. First results show the track-processing system based on our algorithm is feasible, and essential steps are successfully implemented on the real board at the nominal clock speed. The system is capable of reconstructing tracks at a 1.8 MHz event rate, using boards that had originally been designed for 1 MHz readout-only functionality. Moreover, the requested additional hardware to implement the tracking functionality is also comparable with respect to what is needed for the readout-only function. Performances mentioned above are expected to be easily scalable to higher speeds, as much larger and faster FPGA devices are already available today on the commercial market, and another prototype to perform speed test is expected to be built in the next months. Further developments and synergies with fast and smart tracking detector may lead to future experiments with detector-embedded data reconstruction.

References

[1] G.W. Foster, J. Freeman, C. Newman-Holmes, J. Patrick, Nucl. Inst. & Meth. A **269**, 93 (1988)

[2] M. Shochet, L. Tompkins, V. Cavaliere, P. Giannetti, A. Annovi, G. Volpi, Tech. Rep. CERN-LHCC-2013-007, ATLAS-TDR-021, CERN, Geneva (2013), `https://cds.cern.ch/record/1552953`

[3] A. Tapper, D. Acosta (CMS collaboration), Tech. Rep. CERN-LHCC-2013-011. CMS-TDR-12, CERN (2013), `https://cds.cern.ch/record/1556311`

[4] G. Hall, Nucl. Inst. & Meth. A **824**, 292 (2016)

[5] L. Ristori, Nucl. Inst. & Meth. A **453**, 425 (2000)

[6] P. Hough, Proc. Int. Conf. High Energy Accelerators and Instrumentation **C590914** (1959)

[7] P. Hough, US Patent **3069654** (1962)

[8] W.H. Smith, *Trigger and Data Acquisition for hadron colliders at the Energy Frontier* (2013), `arXiv:1307.0706`

[9] A. Abba, F. Bedeschi, M. Citterio, F. Caponio, A. Cusimano, A. Geraci, F. Lionetto, P. Marino, M.J. Morello, N. Neri et al., Tech. Rep. LHCb-PUB-2014-026, CERN-LHCb-PUB-2014-026, CERN, Geneva (2014), `https://cds.cern.ch/record/1667587`

[10] A. Piucci, Master's thesis, University of Pisa, Pisa, Italy (2014), `https://etd.adm.unipi.it/theses/available/etd-06242014-055001/`

[11] B. Angelucci, E. Pedreschi, M. Sozzi, F. Spinella, JINST **7**, C02046 (2012), `http://iopscience.iop.org/article/10.1088/1748-0221/7/02/C02046`

[12] D. Ninci, Master's thesis, University of Pisa, Pisa, Italy (2014), `https://etd.adm.unipi.it/theses/available/etd-11302014-212637/`

Machine Learning and Parallelism in the Reconstruction of LHCb and its Upgrade

Michel **De Cian**[1,a] on behalf of the LHCb collaboration

[1] *Physikalisches Institut der Universität Heidelberg, Germany*

Abstract. The LHCb detector at the LHC is a general purpose detector in the forward region with a focus on reconstructing decays of c- and b-hadrons. For Run II of the LHC, a new trigger strategy with a real-time reconstruction, alignment and calibration was employed. This was made possible by implementing an offline-like track reconstruction in the high level trigger. However, the ever increasing need for a higher throughput and the move to parallelism in the CPU architectures in the last years necessitated the use of vectorization techniques to achieve the desired speed and a more extensive use of machine learning to veto bad events early on. This document discusses selected improvements in computationally expensive parts of the track reconstruction, like the Kalman filter, as well as an improved approach to get rid of fake tracks using fast machine learning techniques. In the last part, a short overview of the track reconstruction challenges for the upgrade of LHCb, is given. Running a fully software-based trigger, a large gain in speed in the reconstruction has to be achieved to cope with the 40 MHz bunch-crossing rate. Two possible approaches for techniques exploiting massive parallelization are discussed.

1 Introduction

For Run II of the LHC, LHCb adopted a new trigger strategy with a real-time reconstruction, alignment and calibration. This allows to directly use the output of the trigger reconstruction to perform physics analyses [1]. To achieve this goal, the track reconstruction in the high-level software trigger (HLT, online) and offline had to be unified, to already obtain the best performance for the real-time alignment and calibration online. Several strategies were employed. A new sequence for the track reconstruction was implemented, which lead to a large speed up. This will be discussed in Section 2. For very computing intensive parts of the code, parallelization in the form of SIMD (single instruction, multiple data) instructions were used that lead to a large gain in time, see Section 3. Furthermore, machine learning methods were employed to reject tracks which do not originate from a real particle (Section 4). Section 5 then discusses the total gain in timing.

The upgraded LHCb detector, which will start taking data in 2020, will use a fully software-based trigger to cope with the 40 MHz bunch crossing rate. In Section 6 two possible strategies and one technical aspect on how to perform track reconstruction in this high multiplicity environment are discussed.

[a] e-mail: michel.de.cian@cern.ch

 DOI: 10.1051/epjconf/201612700006

2 The LHCb detector

The LHCb experiment (see Figure 1) is dedicated to the study of beauty and charm hadron decays with special attention paid to CP violating phenomena as well as searches for physics beyond the Standard Model through rare decays. Since the beauty production cross-section is dominated by gluon fusion, which occurs in the forward region, the LHCb detector is a single-arm forward spectrometer with a pseudorapidity η spanning the range $1.8 < \eta < 4.9$ [2, 3]. The detector includes a high-precision tracking system consisting of a silicon-strip vertex detector (VELO) surrounding the pp interaction region, a large-area silicon-strip detector (TT) located upstream of a dipole magnet with a bending power of about 4 Tm, and three stations of silicon-strip detectors (IT) and straw drift tubes (OT) placed downstream of the magnet. The tracking system provides a measurement of the momentum p of charged particles with a relative uncertainty that varies from 0.5% at low momentum to 1.0% at 200 GeV/c. The minimum distance of a track to a proton-proton collision, the impact parameter, is measured with a resolution of $(15 + 29/p_{\mathrm{T}})\,\mu\mathrm{m}$, where p_{T} is the component of the momentum transverse to the beam, in GeV/c. Different types of charged hadrons are distinguished using information from two ring-imaging Cherenkov detectors (RICH1, RICH2). Photons, electrons and hadrons are identified by a calorimeter system consisting of scintillating-pad (SPD) and preshower detectors (PS), an electromagnetic calorimeter (ECAL) and a hadronic calorimeter (HCAL). Muons are identified by a system composed of alternating layers of iron and multi-wire proportional chambers (MUON).

The LHCb coordinate system is a right handed Cartesian system with the origin at the interaction point. The x-axis is oriented horizontally towards the outside of the LHC ring, the y-axis is pointing upwards with respect to the beam line and the z-axis is aligned with the beam direction.

The online event selection is performed by the trigger, which is composed of three stages: a stage implemented in hardware known as Level 0 (L0) and two stages implemented in software called High-Level-Trigger (HLT1 and HLT2). The HLT1 performs a partial reconstruction of the candidates. In this stage all requirements are inclusive, which means that the selection is applied only to subset of the final state particles. The HLT2 contains both inclusive and exclusive algorithms which are more time-consuming but provide a more precise reconstruction. Exclusive algorithms are used to select a specific decay at the trigger stage.

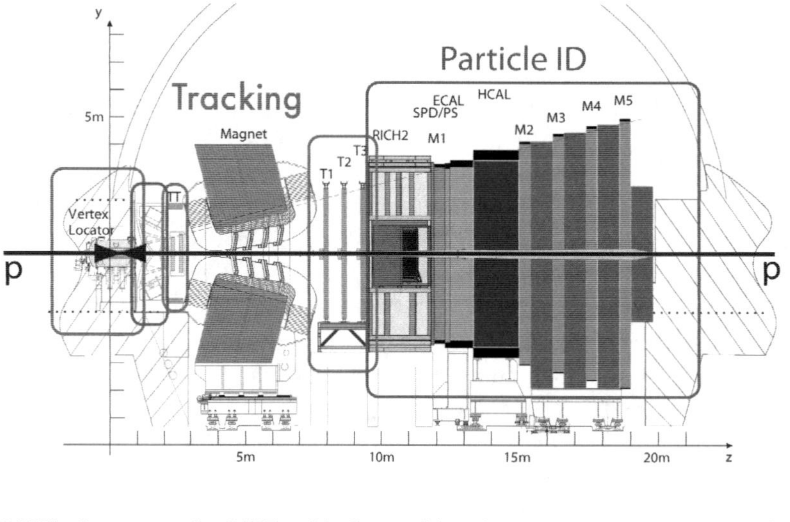

Figure 1. The LHCb detector at the LHC, with the tracking detectors (VELO, TT, T-stations) in red and the detectors for particle identification (RICH1, RICH2, calorimeters and muon system) in blue.

2.1 Track types

With its three tracking detectors, several track types can be reconstructed in LHCb, see Figure 2. The most important category are so-called long tracks, which traverse the full tracking systems and therefore deposit charge in the VELO, the T-stations and potentially also in TT. They are formed from daughters of promptly decaying particles and from daughters of b- and c-hadrons, which only travel a couple of mm before their decay. The other important type are so-called downstream tracks. They originate from daughters of long-living particles, which decay after the VELO and therefore are only reconstructed in TT and the T-stations. Long tracks are reconstructed using two dedicated algorithms. The first, called forward tracking, starts with VELO [4] or upstream tracks [5] and tries to find corresponding clusters in the T-stations [6]. The second, called track matching, uses tracks in the VELO and the T-stations [7] (called T tracks) as starting point and tries to match them in the magnet region [8, 9]. Downstream tracks are reconstructed by using tracks in the T-stations as starting point and extrapolating them back through the magnet [10].

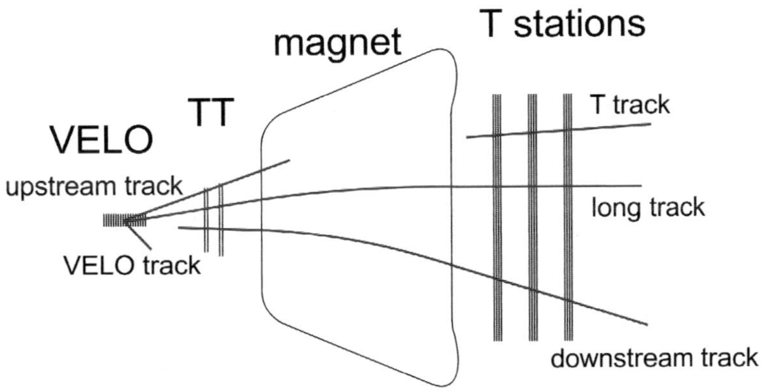

Figure 2. Track types in LHCb. Long tracks and downstream tracks are used for physics analyses, the other types either serve as a component of another track type or are used for detector studies.

2.2 Run II track reconstruction sequence

The run II track reconstruction sequence is split in an HLT1 and HLT2 part (see left part of Figure 3). In HLT1, all VELO tracks are reconstructed. These VELO tracks are fitted with a simplified Kalman filter (allowing for a single scattering at the sensor planes) and are then used to find the primary vertices. The VELO tracks are also extrapolated to TT, where clusters around the extrapolation are searched for. Due to the fringe field of the magnet this provides a first estimate of the sign of the charge. The tracks are then propagated through the magnet to the T-stations, searching for corresponding clusters, such that all tracks with a transverse momentum larger than 500 MeV/c are found (right part of Figure 3). The tracks found are fitted with a Kalman Filter.

Requiring clusters in TT when starting from VELO tracks significantly reduces the time spent in HLT1 to find long tracks, as tracks with low momenta will not have corresponding clusters in TT and therefore will not be propagated through the magnet. Furthermore, having a charge estimate reduces the search window for clusters in the T-stations by a factor of two. And finally, VELO tracks which are fakes will not get confirmed if no clusters in TT are found.

In HLT2, the HLT1 reconstruction part is repeated first. Then all VELO tracks which could not be made into long tracks are again propagated to the T-stations, however without requiring clusters in TT or a cut on the transverse momentum. In addition, T tracks are reconstructed and combined with VELO tracks, to form long tracks, and with clusters in TT, to form downstream tracks. All tracks are fitted with a Kalman filter and clones from the different algorithms are removed.

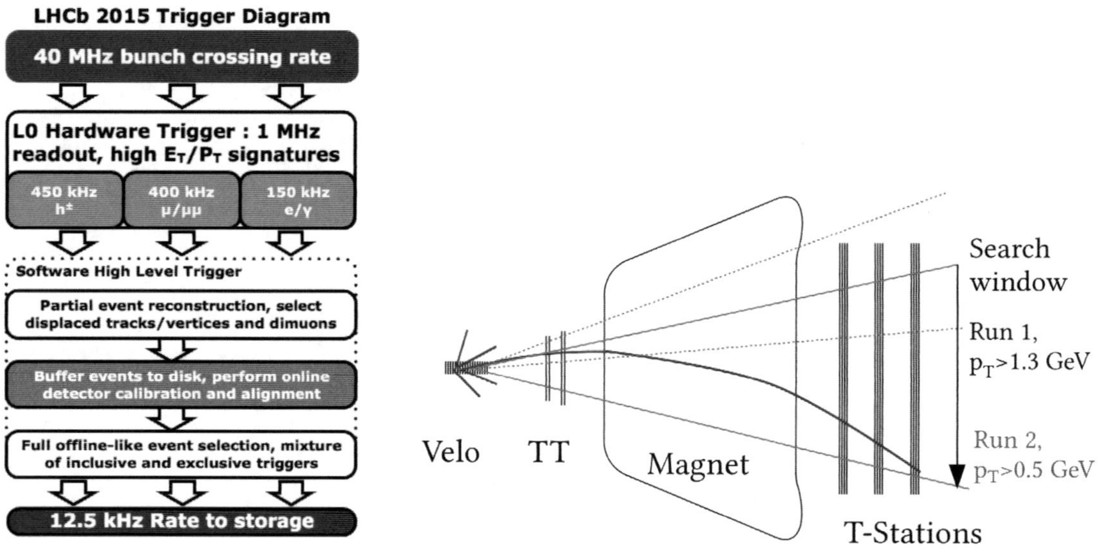

Figure 3. Scheme of the trigger for Run II of LHCb (left). Scheme for the reconstruction of long tracks in Run I and Run II (right). Note that the p_T threshold could be lowered in Run II, allowing for a higher physics yield.

3 SIMD in LHCb

3.1 The need for speed

Even though the new sequence resulted in a reduction in execution time, a further decrease was needed. In order to achieve this, SIMD instructions were introduced. There were two main problems. Most of the code of the LHCb track reconstruction was written several years ago and was not adapted to make use of parallelization natively. Furthermore, a large part of the pattern recognition code consists of selecting the right clusters, tracks, etc., and therefore makes heavy use of branching. SIMD uses vector instructions, meaning that basic mathematical operations like addition and multiplication can be executed in parallel on more than one value. However, this only works if it is guaranteed that all values actually will be computed, and not discarded due to choosing the wrong branch. It is therefore best suited for those parts of the reconstruction code, where purely mathematical computations play a large role in the total time consumption.

SIMD instructions have seen a big evolution in the last 15 years. All CPUs used in the event-filter farm of LHCb can use at least the SSE2 instruction set, which allows processing two double or four float values in parallel. Accordingly, this was the architecture that targeted.

3.2 Hough transform in forward tracking

In the forward algorithm, a VELO (or upstream) track, defining the position and the slopes, is matched to a single cluster in the T-stations. This completely defines all five parameters of the track. All clusters that belong to the same track should therefore accumulate in the same point when projected to a plane parallel to the tracking stations at a given point in z ("Hough plane"), assuming that the track model is correct and neglecting multiple scattering (see Figure 4). This accumulation can then be used to find valid track candidates. The projection of the cluster from its position in the detector plane to the Hough plane uses a third order polynomial, whose coefficients are derived in an iterative way. For each VELO track, all clusters have to be projected exactly once and they are independent of each other, which makes it an ideal case for SIMD instructions.

The overall reduction in time of using SIMD compared to processing the clusters sequentially is about 40%, which is close to the theoretical maximum of 50%, when using SSE2 and double values. The difference stems from the fact that the positions of the clusters are in a layout which needs to be made usable for SIMD first.

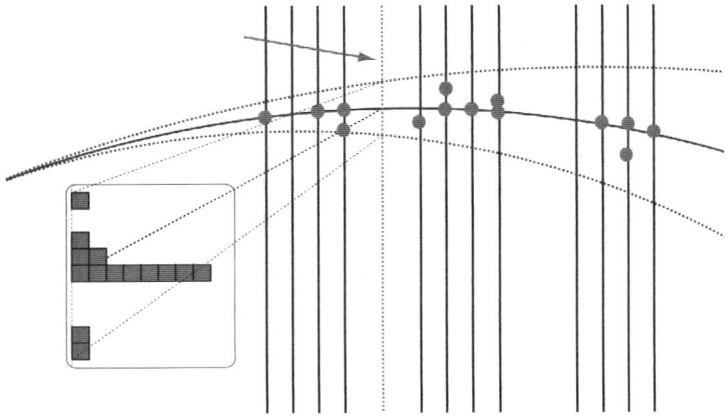

Figure 4. Illustration of the Hough transform. Clusters belonging to the same track (solid blue line) will accumulate around the same point in the Hough plane (red-dashed line) when being projected from their original position. Other clusters will be distributed randomly.

3.3 Computations in the Kalman Filter

All tracks found by the pattern recognition are subsequently fitted with a Kalman filter. The Kalman filter is very time consuming, making it the largest contributor time-wise in HLT1. A significant part of the time is spent in mathematical operations, where one important operation is the propagation of the covariance matrix of the state n to the state $n + 1$. This requires two multiplications of two 4x4 matrices F and C, such as $F \cdot C \cdot F^{\mathrm{T}}$, where C is symmetric. However, a multiplication of two matrices can always be seen as the sum of the multiplications of a row with a single value (see Figure 5), which is ideal for using SIMD instructions. Reordering the multiplications such that they are vectorized reduced the time consumption by about a factor of 2. Using the AVX instruction set (which allows for the parallel processing of 4 double or 8 float values) would lead to a factor of five gain in speed, which is on average two floating point operations per CPU cycle.

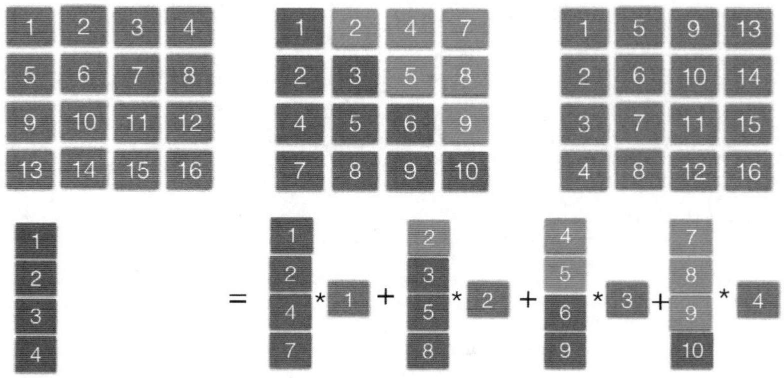

Figure 5. Illustration of the multiplication of three matrices, $F \cdot C \cdot F^{\mathrm{T}}$. The second row shows how the multiplication of the last two matrices can be decomposed in the sum of multiplications of a row-vector with a scalar value, which can be exploited with SIMD instructions.

4 Machine learning

Due to its long lever arm between the tracking stations before and after the magnet, a relatively large fraction of reconstructed tracks are "fake", i.e. they do not correspond to a real charged particle having crossed the detector. In Run I, a cut on the χ^2/ndof of the tracks after the Kalman filter was applied. Later, the user of the data could select tracks with a low fake-probability by cutting on a quantity called "ghost probability". This classifier was based on a neural net.

Approximately 22% of all reconstructed long tracks after the cut on χ^2/ndof were estimated to be fake in Run II. To further reduce this rate in the reconstruction, the idea was to use the ghost probability classifier directly in the final selection of tracks .

For Run II, the ghost probability is still based on a neural net, however it only uses variables as input which do not consume much computing power in the calculation. In total, 21 input variables are used. Most of them are taken from the result of the Kalman filter, like the overall χ^2/ndof, but also the χ^2/ndof of segments of the track, for example the TT – T-station segment. These segments might be more sensitive to detecting so called matching ghosts. They occur when a real VELO or upstream track and a real track in the T-stations are matched, although originating from different particles. The neural net consists of one hidden layer with 26 nodes. No gain in performance was seen by training a deep neural net. Likewise no gain was seen by using human assisted learning, i.e. forming new variables from the raw quantities as an input to the neural net.

To further speed up the calculation, the activation function of the neural net was chosen to be $1/\sqrt{1+x^2}$. This decreased the time consumption by a factor of 50 compared to using the hyperbolic tangent as activation function, which was the default in the Run I classifier.

All these improvements lead to an average time of the calculation of the ghost probability of 0.5ms per event; it therefore became possible to evaluate it in the high level trigger and to use it as a track-quality variable. The new scheme applies a combined cut on χ^2/ndof and ghost probability, which reduces the rate of fake tracks from 22% to 14%, without any loss of real tracks. An illustration of applying the ghost probability on tracks from $D^0 \rightarrow K\pi$ and $K_S \rightarrow \pi\pi$ decays can be seen in Figure 6.

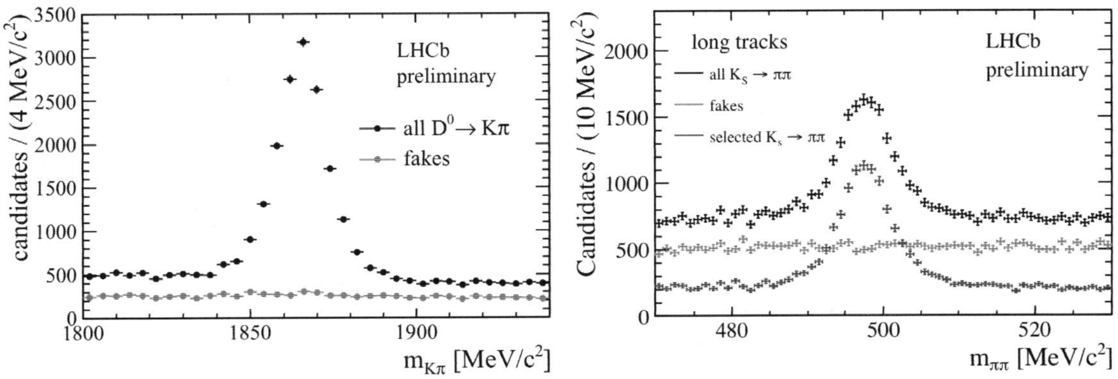

Figure 6. Performance of the ghost probability classifier for long tracks and $D^0 \to K\pi$ (left) and $K_S \to \pi\pi$ (right) decays. The rejected events are shown in red and show no peaking structure, meaning that no signal is lost with the cut on the ghost probability.

5 Total gain in timing

In addition to the presented points, many small-scale optimisations were performed in all the pattern recognition algorithms and in the Kalman filter, for example more efficient calculations or better use of the C++11 standard. The total speed-up, when comparing the reconstruction sequence in Run I and Run II, was a factor of 2, without any loss in performance. To achieve the same physics results without any modification of the code would have resulted in buying more machines for the HLT farm, whose total cost would have been several million Swiss francs.

6 Track reconstruction in the upgrade of LHCb

The LHCb detector will be upgraded in the long shutdown 2 of the LHC and restart taking data in 2020 at an instantaneous luminosity of $2 \cdot 10^{33}\,\mathrm{cm}^{-2}\,\mathrm{s}^{-1}$, which is about a factor of 5 more than the current instantaneous luminosity. At this luminosity, the rate of potentially interesting physics events would be too high for the L0 hardware trigger to handle. The idea is therefore to have only a software trigger which runs at the full bunch-crossing frequency. In order to achieve this goal, the tracking system needs to be replaced. The current VELO, a silicon strip detector, will be replaced by a pixel-based solution with a very high track reconstruction efficiency [11]. The TT-station before the magnet will be replaced by a silicon strip detector with a fine granularity [12]. The tracking stations behind the magnet will consist of 12 layers of a scintillating fiber tracker, which is read out by silicon photomultipliers [12].

The overall time available for track reconstruction is expected to be about 15 ms per event. This places strong constraints on all pattern recognition algorithms and the track fitter. The only way to achieve this goal is to exploit parallelisation, either using CPUs, or using other architectures, for example GPUs or Intel Xeon Phi. The main question is not only about the best raw performance, but mainly about the best performance for the amount of invested money.

6.1 Cellular automaton tracking

The search for standalone tracks in the T-stations is a highly parallelizable problem. For a given station (with its four layers in an *x*-*u*-*v*-*x* arrangement), for each cluster in a given layer, corresponding clusters have to be found in the other layers. One idea to perform this track reconstruction is to use a cellular automaton approach (see Ref. [13] for an example). First, tracklets between the two x layers in all three T-stations are reconstructed. Then stereo hits are added, leading to so-called "stubs". These stubs now need to be connected between the three stations: Three-stations stubs are preferred over two-station stubs, as the latter are more likely to be fake tracks.

To test the efficiency of this approach (irrespective of the time consumption), a CPU implementation was written for the Run II detector. The track reconstruction efficiency of the cellular automaton approach is within a couple of percent of the nominal algorithm to reconstruct T tracks, where it has to be taken into account that the Run II algorithm has undergone extensive tuning over several years. The next step is now to port the algorithm to a GPU to test the performance not only for track reconstruction efficiency, but mainly for time consumption.

6.2 VELO tracking with GPUs

Another approach to gain experience with GPUs is to take an existing track reconstruction algorithm (in this case for the VELO standalone tracking), and to port it to a GPU, using the CUDA language [14]. This algorithm runs in "parasitic" mode in the monitoring farm of the HLT, meaning that it can be tested in real data taking conditions. While the efficiencies between the CPU and the GPU implementations are very similar, the GPU, as expected, shows an advantage when several events can be prepared in such a way that they can be processed in parallel, see Figure 7.

6.3 Structure of arrays

In order to optimally exploit SIMD instructions, the data needs to be provided in a "vectorization-friendly" way. At the moment, most classes in LHCb are arranged in an array-of-structures. For example, a cluster in a tracking detector is represented (in a simplified way) with a class "cluster" and holds three members, its coordinates x, y, z. A set of clusters therefore is arranged as an array of the cluster class. However, when using SIMD instructions, for example to calculate something with the x-coordinates only, the values are not sequential in memory. This prevents SIMD to process them simultaneously. The solution is to arrange the clusters in three arrays, each of them holding one coordinate only. This is less intuitive, but accessing the x-coordinates would mean accessing values which are next to each other in memory, greatly benefitting vectorization with SIMD.

Studies have started in LHCb that exploit the structure-of-array arrangement to speed up critical parts of the code.

7 Conclusion

For Run II, LHCb moved to a new trigger strategy with a real-time reconstruction, alignment and calibration. To fulfill the timing constraints, the track reconstruction was greatly sped up using SIMD instructions to increase the speed of the code and machine learning to efficiently remove fake tracks already in the reconstruction. Overall a gain of a factor of two in timing was obtained. The need for speed will continue in the upgrade of the LHCb detector, where a hardware trigger-less readout will necessitate the use of even faster algorithms, and therefore even more parallelized systems.

 DOI: 10.1051/epjconf/201612700006

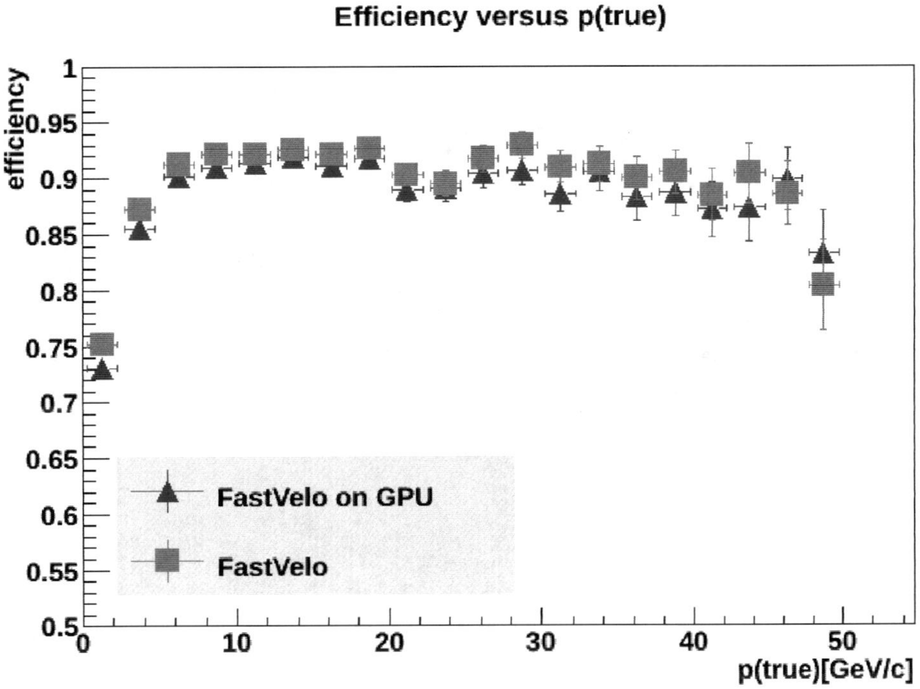

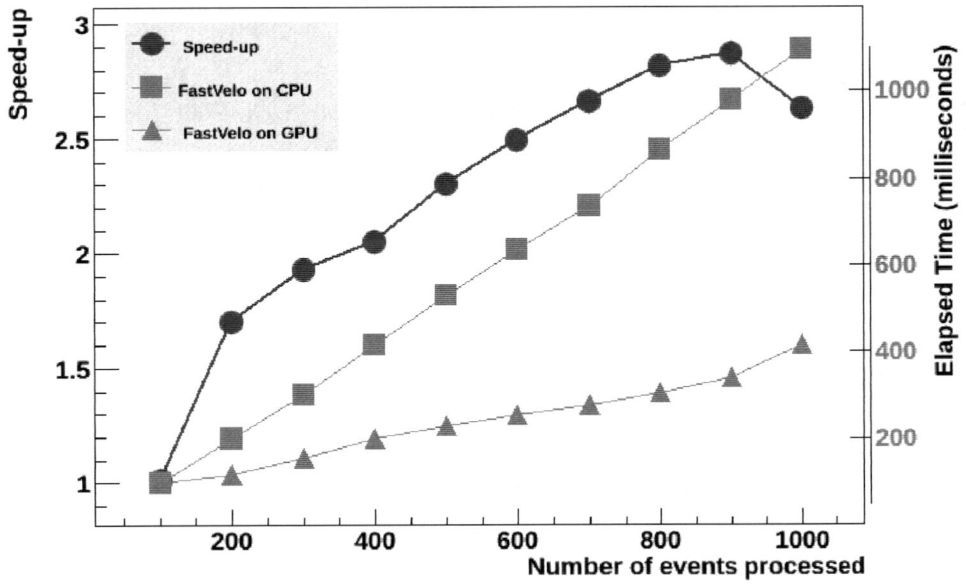

Figure 7. Left: Track reconstruction efficiency of the CPU based FastVELO algorithm (red) and its GPU implementation (blue) as a function of the momentum of the track. Right: Time consumption by the CPU (rectangle) and GPU (triangle) based FastVELO algorithm as a function of the number of processed events.

References

[1] Aaij, R. et al (2016), arXiv:1604.05596

[2] Alves, A. et al., JINST **3**, S08005 (2008)

[3] Aaij R. et al, Int. J. Mod. Phys. A **30**, 1530022 (2015), http://www.worldscientific.com/doi/pdf/10.1142/S0217751X15300227

[4] O. Callot, Tech. Rep. LHCb-PUB-2011-001. CERN-LHCb-PUB-2011-001, CERN, Geneva (2011), https://cds.cern.ch/record/1322644

[5] E.E. Bowen, B. Storaci, M. Tresch, Tech. Rep. LHCb-PUB-2015-024. CERN-LHCb-PUB-2015-024. LHCb-INT-2014-040, CERN, Geneva (2016), https://cds.cern.ch/record/2105078

[6] O. Callot, S. Hansmann-Menzemer, Tech. Rep. LHCb-2007-015. CERN-LHCb-2007-015, CERN, Geneva (2007), https://cds.cern.ch/record/1033584

[7] O. Callot, M. Schiller, Tech. Rep. LHCb-2008-042. CERN-LHCb-2008-042, CERN, Geneva (2008), https://cds.cern.ch/record/1119095

[8] M. Needham, J. Van Tilburg, Tech. Rep. LHCb-2007-020. CERN-LHCb-2007-020, CERN, Geneva (2007), https://cds.cern.ch/record/1020304

[9] M. Needham, Tech. Rep. LHCb-2007-129. CERN-LHCb-2007-129, CERN, Geneva (2007), https://cds.cern.ch/record/1060807

[10] O. Callot, Tech. Rep. LHCb-2007-026. CERN-LHCb-2007-026, CERN, Geneva (2007), https://cds.cern.ch/record/1025827

[11] LHCb Collaboration, Tech. Rep. CERN-LHCC-2013-021. LHCB-TDR-013, CERN, Geneva (2013), https://cds.cern.ch/record/1624070

[12] LHCb Collaboration, Tech. Rep. CERN-LHCC-2014-001. LHCB-TDR-015, CERN, Geneva (2014), https://cds.cern.ch/record/1647400

[13] I.S. Kulakov, S.A. Baginyan, V.V. Ivanov, P.I. Kisel, Phys. Part. Nuclei Letters **10**, 162 (2013)

[14] A. Badalov et al., Tech. Rep. LHCb-PUB-2014-034. CERN-LHCb-PUB-2014-034, CERN, Geneva (2014), https://cds.cern.ch/record/1698101

EPJ Web of Conferences **127**, 00007 (2016) DOI: 10.1051/epjconf/201612700007
Connecting The Dots 2016

Full Offline Reconstruction in Real Time with the LHCb Detector

Agnieszka **Dziurda**[1,a] on behalf of the LHCb Collaboration

[1] *CERN, Geneva, Switzerland*

Abstract. This document describes the novel, unique in High Energy Physics, real-time alignment and calibration of the full LHCb detector. The LHCb experiment has been designed as a dedicated heavy flavour physics experiment focused on the reconstruction of c and b hadrons. The LHCb detector is a single-arm forward spectrometer, which measures proton-proton interactions at the LHC. The operational bunch crossing rate is several orders of magnitude above the current abilities of data recording and storage. Therefore, a trigger system has been implemented to reduce this rate to an acceptable value. The LHCb trigger system has been redesigned during the 2013–2015 long shutdown, achieving offline-quality alignment and calibration online. It also allows analyses to be performed entirely at the trigger level. In addition, having the best performing reconstruction in the trigger gives the possibility to fully use the particle identification selection criteria and greatly increases the efficiency, in particular for the selection of charm and strange hadron decays. This results in enhancing the physics programme of the LHCb experiment.

1 Introduction

The LHCb reconstruction and trigger are sensitive to the LHC running conditions. As the LHC increased its energy from the $\sqrt{s}$ = 7(8) TeV in 2011 (2012) to $\sqrt{s}$ = 13 TeV for Run II, (2015-2018), the strategy of data taking had to be redesigned and reoptimized. Additionally, the performance of a large computing cluster used for the software trigger and known as the Event Filter Farm (EFF) roughly doubled between 2012 and 2015, which significantly increased the time budget for the online reconstruction. This allowed an improved reconstruction, and coherent real-time detector alignment and calibration to be performed inside the trigger.

2 The LHCb detector

The LHCb experiment is dedicated to the study of beauty and charm hadron decays with special attention paid to CP violating phenomena as well as searches for physics beyond the Standard Model through rare decays. Since the beauty production cross-section is dominated by gluon fusion, which more likely occurs in the forward region, the LHCb detector was designed as a single-arm forward spectrometer with a pseudorapidity η spanning the range $1.8 < \eta < 4.9$ [1, 2].

[a] e-mail: agnieszka.dziurda@cern.ch

 DOI: 10.1051/epjconf/201612700007

A schematic view of the LHCb detector is shown in Figure 1. The detector includes a high-precision tracking system consisting of a silicon-strip vertex detector (VELO) surrounding the pp interaction region, a large-area silicon-strip detector (TT) located upstream of a dipole magnet with a bending power of about 4 Tm, and three stations (T1–T3) of silicon-strip detectors (IT) and straw drift tubes (OT) placed downstream of the magnet. The tracking system provides a measurement of the momentum p of charged particles with a relative uncertainty that varies from 0.5% at low momentum to 1.0% at 200 GeV/c. The minimum distance of a track to a proton-proton collision, the impact parameter, is measured with a resolution of $(15+29/p_T)$ μm, where p_T is the component of the particle momentum transverse to the beam, in GeV/c. Different types of charged hadrons are distinguished using information from two ring-imaging Cherenkov detectors (RICH1, RICH2). Photons, electrons and hadrons are identified by a calorimeter system consisting of scintillating-pad (SPD) and preshower detectors (PS), an electromagnetic calorimeter (ECAL) and a hadronic calorimeter (HCAL). Muons are identified by a system composed of alternating layers of iron and multi-wire proportional chambers (M1–M5).

The LHCb coordinate system is a right handed Cartesian system with the origin at the interaction point. The x-axis is oriented horizontally towards the outside of the LHC ring, the y-axis is pointing upwards with respect to the beam line and the z-axis is aligned with the beam direction.

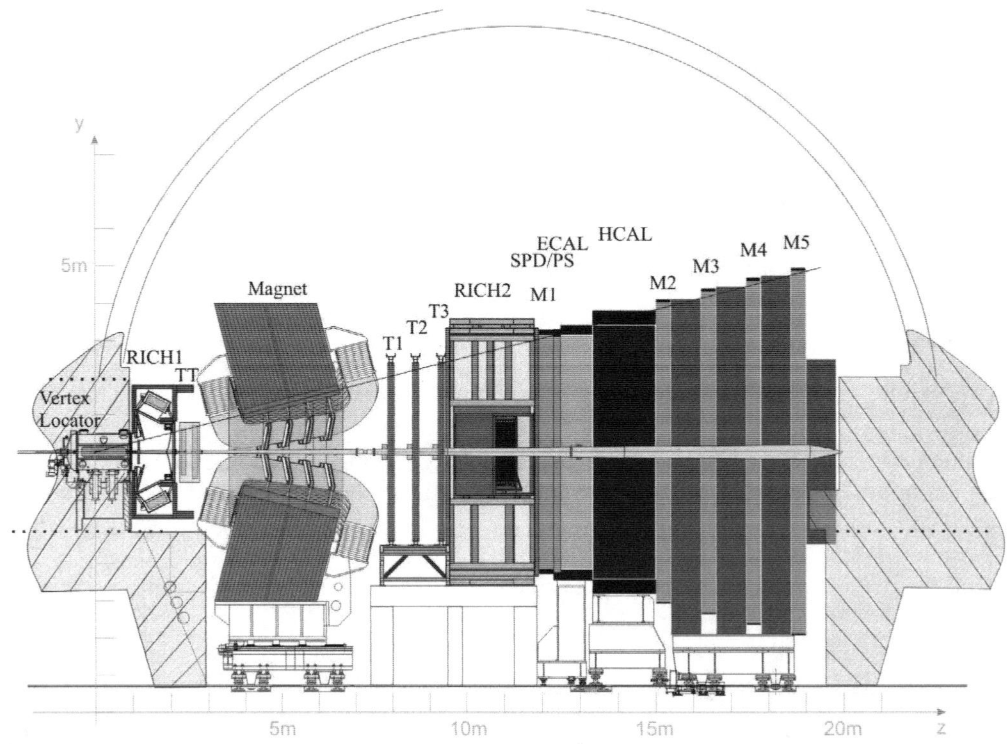

Figure 1. A schematic view of the LHCb detector. The Vertex Locator indicates the vertex detector. The labels RICH1 and RICH2 denote the Ring Imaging Cherenkov detectors, TT and T1–T3 are the tracking stations, the latter is composed of IT and OT detectors, M1–M5 indicates the muon chambers, SPD (PS) is the scintillator pad (preshower) detector. Finally ECAL and HCAL mark the electromagnetic and hadronic calorimeters, respectively.

The online event selection is performed by the trigger, which is composed of three stages: one hardware stage, known as Level0 (L0), and two software stages called High-Level-Trigger (HLT1 and HLT2). The HLT1 trigger performs a partial reconstruction of the candidates from many decays. At this stage all requirements are inclusive, which means that the selection is applied only to a subset of the final state particles. The HLT2 trigger contains both inclusive and exclusive algorithms which are more time-consuming but provide a more precise reconstruction. Exclusive algorithms are used to select a specific decay.

3 Strategy of data taking

The LHC running conditions have significantly changed between Run I and Run II. To meet the challenge due to increasing energy (from 7(8) TeV to 13 TeV) and the operational bunch crossing rate (from 15 to 30 MHz), the strategy of data taking has been redesigned during the 2013-2015 long shutdown. The new strategy results in an increase of the output rate from 5 kHz (Run I) to 12.5 kHz (Run II). Detailed information about the performance of the trigger in Run I can be found in [3].

The schematics of the DAQ strategies in Run I and II are shown in Figure 2. The first stage, knows as the low level trigger (L0), is common for both periods of data taking. The main purpose of the system is to quickly find general signatures of interesting decays and reduce the rate of accepted candidates to about 1 MHz, at which the full detector can be read out. This hardware trigger is based on information from the calorimeter system and muon chambers.

All candidates accepted by the L0 trigger are processed by the software trigger, which runs on the EFF. The farm currently consists of approximately 1700 nodes (800 of them were added for Run II) with 27000 physical cores. During Run I limited resources allowed only a simplified reconstruction to be performed inside the HLT. In particular, faster but less accurate primary vertex and track reconstructions were performed and only partial particle identification information was accessible. Moreover preliminary calibration and alignment were used. After data taking all selected events were processed further through an offline reconstruction together with updated calibration and alignment parameters, resulting in the best performing reconstruction. As a consequence significant differences between online and offline data quality were observed.

Due to operational issues in 2012, the LHC delivered stable beam conditions for approximately 30% of the scheduled time. In the LHCb trigger system about 20% of the L0 accepted events were temporarily buffered on the EFF nodes and processed in the gaps between stable beam periods. This deferred triggering method allowed LHCb to increase the data sample for physics analysis. During Run II the HLT1 and HLT2 stages are split. All events passing HLT1 are buffered on the EFF farm, which doubles the effective processing power. Thanks to this approach further HLT2 event processing can be done during gaps. In addition, this offers the needed time for the run-by-run evaluation of the detector calibration and alignment. Thus the HLT2 provides a full event reconstruction at optimal offline quality.

4 Real-time alignment and calibration

The detector alignment and calibration vary due to pressure and/or temperature changes, movements caused by operational conditions such as, for instance, switching the magnetic field direction or mechanical interventions. Thus updates of alignment and calibration constants are crucial to achieve optimal physics performance. The correct alignment of the VELO is needed for an accurate identification of primary and secondary vertices, which is visible in impact parameter, primary vertex (PV) and decay-time resolutions. In addition, more precise alignment of the entire tracking system

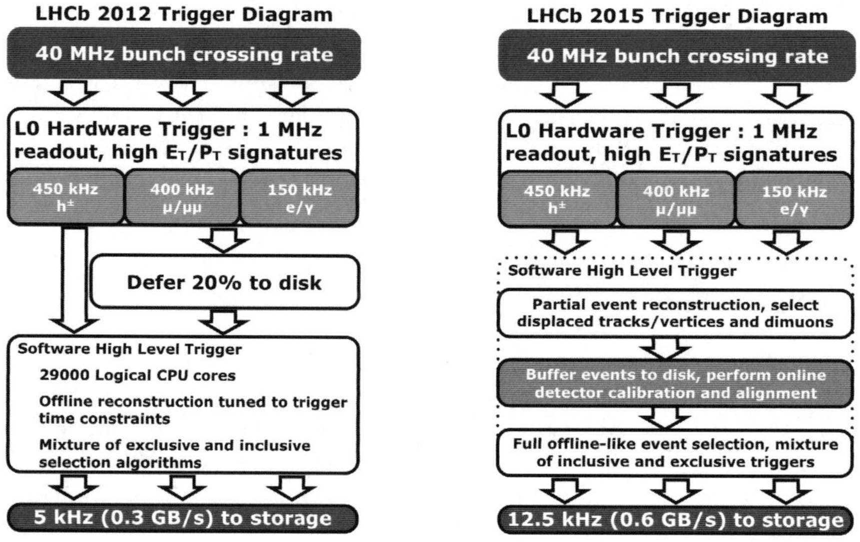

Figure 2. The layout of the LHCb trigger for Run I (left) and Run II (right) data taking. Software High Level Trigger indicates HLT1 and HLT2 stages.

directly implies better mass resolution, as shown in Figure 3. The Υ invariant mass resolution is significantly improved from $86\,\mathrm{MeV}/c^2$ for the very preliminary constants up to $44.3\,\mathrm{MeV}/c^2$ for the last alignment. Finally the precise calibration and alignment of RICH detectors implies better particle identification and allows to increase the purity of the sample. To ensure the best physics performance, the position and orientation of detector elements in the global reference frame must be known with an accuracy significantly better than the single hit resolution.

The main alignment and calibration tasks have dedicated HLT1 trigger lines, which collect certain types of events needed by each task, for instance $D^0 \to K\pi$, $J/\psi \to \mu\mu$, or tracks with specific properties. The selected events are saved to the disk, and reconstructed within the EFF. The alignment

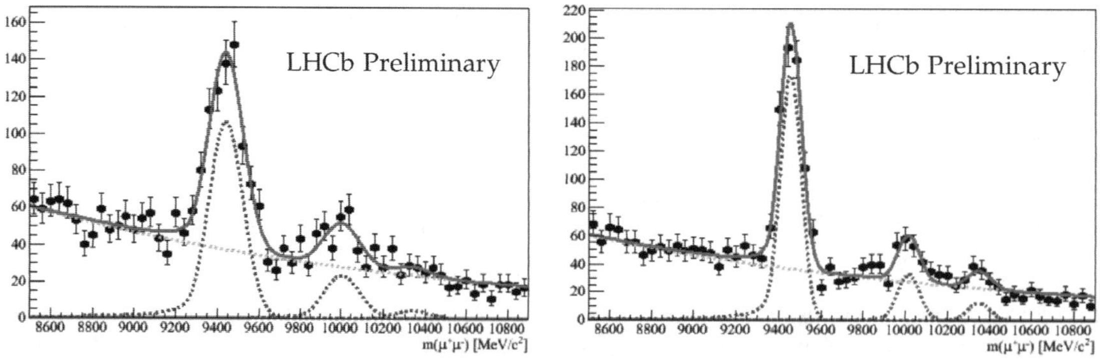

Figure 3. The Υ invariant mass resolution with very first (left) and very last (right) tracking alignment.

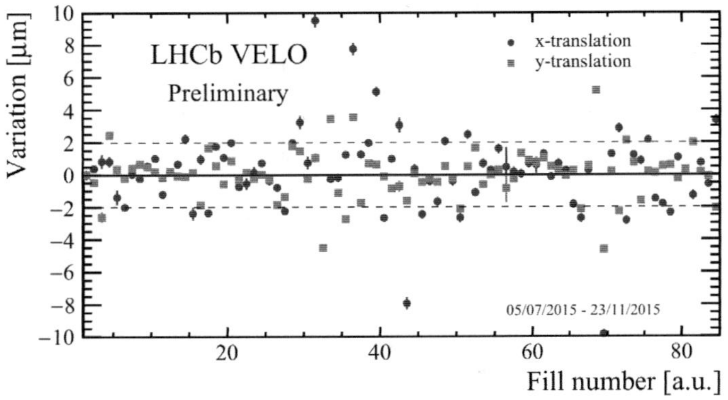

Figure 4. The stability of the VELO half alignment during 2015 data taking. The red (blue) points correspond to x (y) translations, while two horizontal lines denote thresholds within the alignment is not updated. Each point indicates difference of the alignment constants with respect to the value from previous alignment.

and calibrations are performed at regular intervals, which can be at the beginning of the run, fill or less frequently. The resulting constants are updated only if they differ significantly from the currently used values.

4.1 Detector alignment

The tracking alignment is based on an iterative procedure [4, 5], which minimizes the residuals of a Kalman filter and takes into account many effects such as multiple scattering, energy loss in the material, and magnetic field information. In addition, to avoid global distortions, tracks can be combined in vertices, allowing for the use of primary vertex and mass constraints. After the VELO alignment, the algorithm can align all tracking detectors simultaneously.

4.1.1 VELO alignment

The VELO is composed of left and right parts, therefore its two halves can be moved apart during the beam injection period for the safety of the detector and moved back together once the beams are stable. In nominal working conditions the sensitive parts of the VELO sensors are about 8 mm away from the beam. The VELO movement is achieved using stepper motors. The position is read from resolvers mounted on the motor axes with an accuracy better than 10 μm. The automated procedure closes the VELO symmetrically around the beam. In order not to degrade the impact parameter or decay-time resolutions, the VELO sensors need to be aligned with a precision of a few microns in x and y and a few tens of microns in z.

Since the VELO halves are closed for every fill, the precise alignment of this detector for each fill is crucial for the best physics performance. Figure 4 shows the stability of the VELO half alignment for the x and y transitions during 2015 data taking, where the horizontal dashed lines indicate the threshold within the alignment is not updated. Each point denotes the difference between new constants and parameters from the previous alignment. As expected, the constants are updated each few fills.

 DOI: 10.1051/epjconf/201612700007

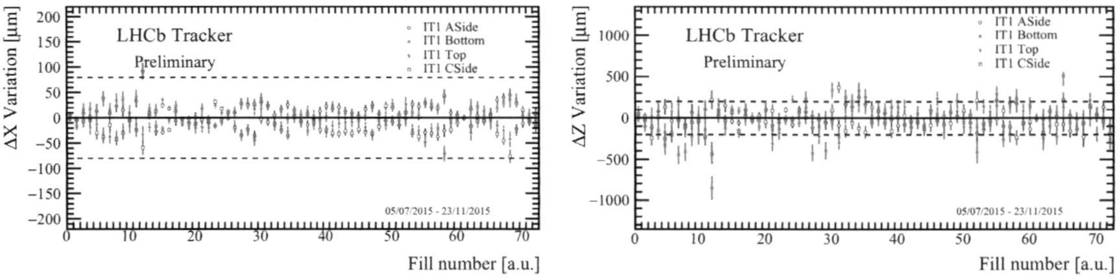

Figure 5. The stability of the IT box alignment for different parts of the detector during 2015 data taking. Two horizontal lines denote the threshold within the constants are not updated. The left plot shows variations for the x direction, while the right corresponds to z translations. Each point indicates the difference of the alignment constants with respect to the value from previous alignment.

4.1.2 Tracking alignment

The rest of the LHCb tracking system is aligned relative to the VELO using tracks. This alignment benefits from the possibility to use mass constraints [4] in the evaluation of the alignment constants. This technique increases the precision of the method. Thus a sample of 100,000 $D^0 \to K\pi$ decays is used as input of this procedure. The alignment is performed for all tracking stations: IT, TT and OT. Typical alignment degrees of freedom are: displacements in x direction, rotations around the z-axis for the smallest structures (such as modules in the TT and boxes in the IT) as well as displacements in z for the layers. Detectors may be moved due to maintenance during accelerator technical stops, which occurs approximately once per two months. In addition, the magnetic field direction is switched about twice per month. It is observed that this causes a significant movement of the IT boxes and requires a few days for the position of these elements to stabilize. As a consequence, to account for any changes in detector conditions, the alignment is performed for every fill, while it is updated when the variations are significant (i.e. above threshold). Figure 5 shows the stability of IT detector during 2015 data taking using the online per fill alignment approach. As before, the horizontal lines denote the threshold between which the constants are not updated. As can be seen the variations are mostly smaller than the precision of the alignment, showing that infrequent updates are sufficient.

4.1.3 Muon stations alignment

Muon detection is performed in the five stations of muon chambers placed in the outermost layers of the LHCb spectrometer. The first station (M1) is located in front of the calorimeter system, while the remaining four stations (M2–M5) are placed behind the system (see Figure 1). The read-out pads of the muon chambers are less fine-grained than other components of tracking system, and as a consequence the spatial resolution of the muon system is less accurate. Therefore, the precise alignment of muon stations is less critical. Nevertheless, the information from the muon system is used in the L0 trigger, where the momentum of muons passing the detector is estimated from information provided by the first two stations. Simulation studies show that an alignment precision better than 1 mm is required to guarantee high trigger efficiency without any significant charge asymmetry. The alignment of the muon detector is required after each mechanical intervention that moves one or both halves of the muon chamber. If no intervention happens, the position is stable. Nevertheless, due to the importance of this alignment for the hardware trigger performance, the alignment of muon chambers is performed online for monitoring purposes. Figure 6 shows the stability of muon stations during 2015 data taking. The position of the muon chambers are stable, with a maximal variation of $O(200\,\mu\text{m})$, much smaller than needed for the optimal performance. Thus the alignment is updated only after hardware interventions.

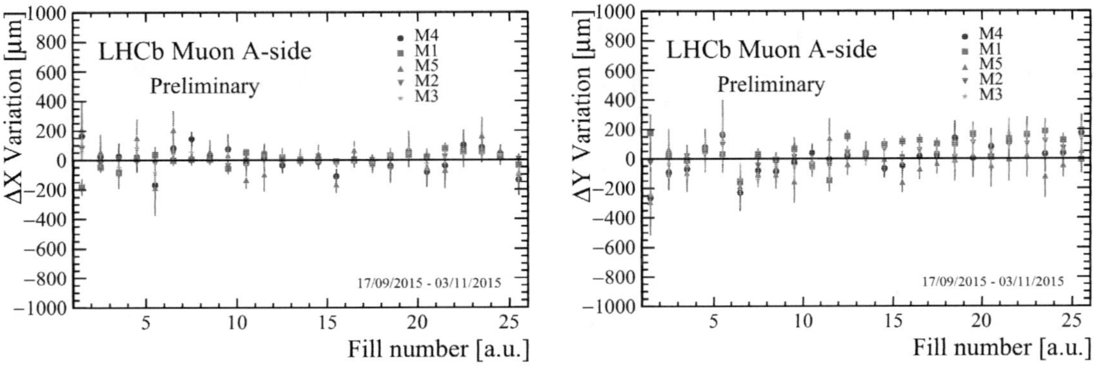

Figure 6. Stability of the muon chambers alignment; M1–M5 denote the different muon stations. The left plot shows the translation along x, while the right one shows the y translations. Each point indicates the difference of the alignment constants with respect to the value from the previous alignment.

4.1.4 RICH alignment

Each of the RICH detectors is composed of spherical and flat mirrors. The light created by particles passing through the detectors is reflected off a primary mirror onto a secondary mirror, and then focused on the photon detection plane. Any misalignment of these mirrors causes the Cherenkov ring on the photodetector plane to be shifted with respect to its expected position. As a consequence, a projected track coordinate is not at the center of the ring. The deviation from the Cherenkov opening angle θ can be parametrized as a function of the azimuthal angle ϕ in the photodetector plane:

$$\Delta\theta(\phi) = T_x \cos(\phi) + T_y \sin(\phi) \tag{1}$$

where T_x and T_y are the components of the misalignment of the projected track coordinate with respect to the expected position, and θ is the Cherenkov angle calculated from the momentum of the track and the refractive index of the radiator. The alignment is determined for each mirror by fitting Equation (1) to the histogram of the $\Delta\theta(\phi)$, as it is presented in Fig 7. The procedure is iterated until the variations are below a certain threshold. Both RICH detectors contain 110 mirrors to be aligned, which results into 1090 constants. The procedure is used in the monitoring system and runs every few fills, taking about one hour to converge.

4.2 Calibration

In addition to alignment, other detector constants have to be calibrated, in particular those of the OT, RICH detectors and calorimeters. A brief description of these calibrations is provided in the following subsections.

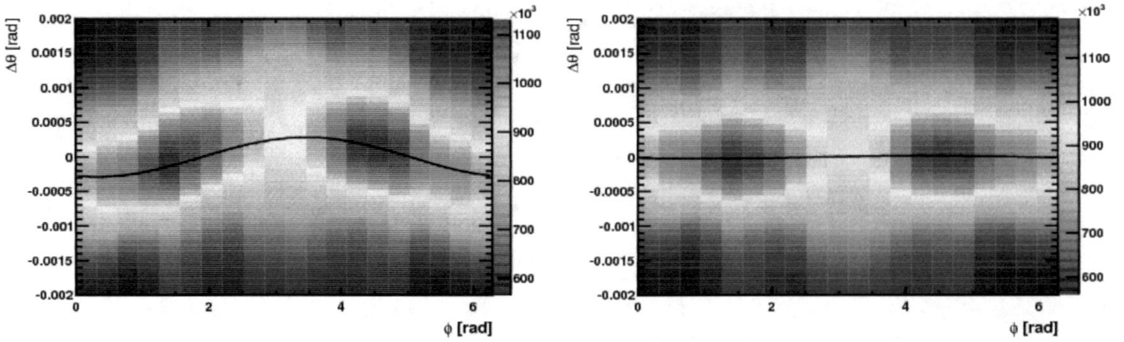

Figure 7. Difference between the measured and expected Cherenkov angle $\Delta\theta$ as a function of the azimuthal angle ϕ before (left) and after (right) the mirror alignment for one mirror.

4.2.1 Outer Tracker calibration

The Outer Tracker is a gaseous straw-tube detector filled with a gas mixture. The straws have an inner diameter of 5 mm. For optimal spatial resolution, the straw layers are shifted by half a straw diameter inside one module. The time difference between a bunch crossing and the measurement of a signal in one of the straws t_{meas} is composed of several components, including the global time offset t_0. A global time offset occurs due to the synchronization between the collision time and the LHCb clock and is common for all modules. A shift of 0.5 ns leads to a tracking inefficiency about 0.25%. The parameter in question is evaluated by fitting the residuals of the reconstructed and expected drift time. As it can be seen in Figure 8 the module time calibration is stable in time.

4.2.2 RICH calibration

The RICH automatic calibration is evaluated and updated in every run. The procedure implements two main calibrations: the RICH refractive index and the Hybrid Photon Detectors (HPDs) images. The refractive index of the gas radiators is sensitive to changes of temperature and pressure, as well

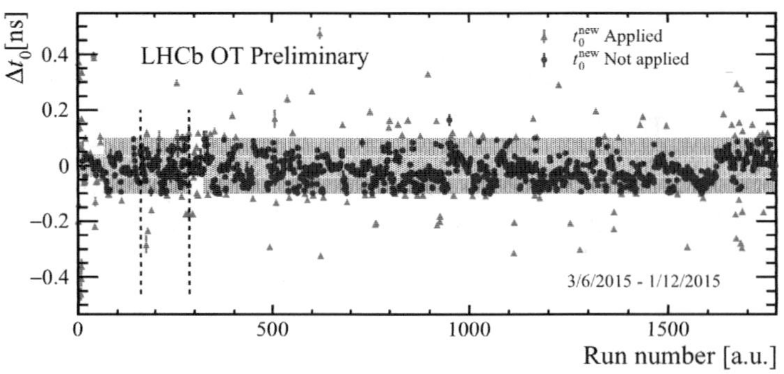

Figure 8. Variation of global time offset Δt_0 during 2015 year of data taking. The shade area together with blue points denote the values within threshold, the red points indicate when the calibration was updated.

as to the exact composition of the gas mixture. All of these components may change in time, thus two corrections parameters, one for each detector, have to be evaluated. They are determined by fitting the difference between the reconstructed and expected Cherenkov angle. In addition, the HPDs may be affected by changes in the magnetic and electrical fields. The calibration consists of 1940 parameters and is obtained by fitting the cleaned anode images using a Sobel filter. Here, the determination of constants is not iterative, and can be determined by fitting the relevant distribution on a single node.

4.2.3 Calorimeters calibration

The LHCb calorimeter system consists of a scintillating-pad detector (SPD), preshower (PS) detectors, and two main calorimeters: an electromagnetic (ECAL) and a hadronic (HCAL) calorimeter (see Figure 1). The system uses two types of the calorimeter gain calibrations: absolute and relative. The absolute calibrations are performed either each month (for ECAL) or in each technical stop (for HCAL), while the relative one runs online every fill.

The absolute energy calibration of the ECAL uses the $\pi^0 \rightarrow \gamma\gamma$ decay, which is found by combining a photon hitting the cell to be calibrated with another reconstructed photon. Then, a fit to the invariant mass $m(\gamma\gamma)$ is performed. The procedure is repeated iteratively. The improvement in the invariant mass resolution between miscalibrated and perfectly calibrated samples is a factor of about 2, from around 18 MeV/c^2 before to about 8 MeV/c^2 after the calibration. In order to apply the procedure to every cell in the ECAL, about 50 million events are required.

The absolute calibration of the HCAL uses two ^{137}Cs sources, one per detector side. A hydraulic system transports the source to all of the scintillator cells, then the response of the photomultiplier tubes (PMTs) is measured, which allows the HCAL parameters to be set.

Since the rate of the L0 trigger strongly depends on the calorimeter gain, it is crucial to keep it constant in time. The time period between absolute calibrations is too long with respect to the gradual gain reduction, thus relative methods of calibrations are used. The response to LED flashes is found to agree well with the response of particles passing the detector, so for monitoring purposes a dedicated LED system was built. The system detects aging of the PMTs, calculates the required adjustment and automatically updates the high voltage in each fill. In addition, to monitor the system it is possible to use the E/pc distribution of electrons (ECAL) and hadrons (HCAL). The variation of the gain can be determined from the variation of the occupancies ratio, where the occupancy is defined as the fraction of events in which the ADC output is above a threshold. The occupancies ratio is checked with respect to a reference sample and is proportional to the changes in gain. When a significant difference is detected, the high voltage is changed accordingly.

5 Reconstruction and performance

For 2015, the time budget of the software trigger was set to be 35 ms for HLT1 and 650 ms for HLT2. The optimal reconstruction had to match these time constraints, so many reconstruction algorithms were reoptimized. A big effort was made in terms of code optimization, especially performing faster approximations, applying vectorization or even revisiting entire algorithms. All of this has been done without losing any performance.

It is worth highlighting some examples. The reconstruction of the pp interaction point (primary vertex) has been reoptimized, resulting in an improved reconstruction efficiency and PV resolution as for the offline processing. The global efficiency increased by about 3% with a reduction of fake primary vertices of 70%. In particular, a significant gain is achieved for the PVs with a low number of associated tracks. Figure 9 shows the performance plots of the tracking system. The tracking efficiency (left) and the resolution of the impact parameter (right) are compared to that found in 2012 data, finding a good agreement. However, it is worth to note that 2012 references were performed as offline reconstruction, while the entire 2015 reconstruction is performed online. Thus, these results show that the track reconstruction in the trigger has achieved offline quality in the year 2015.

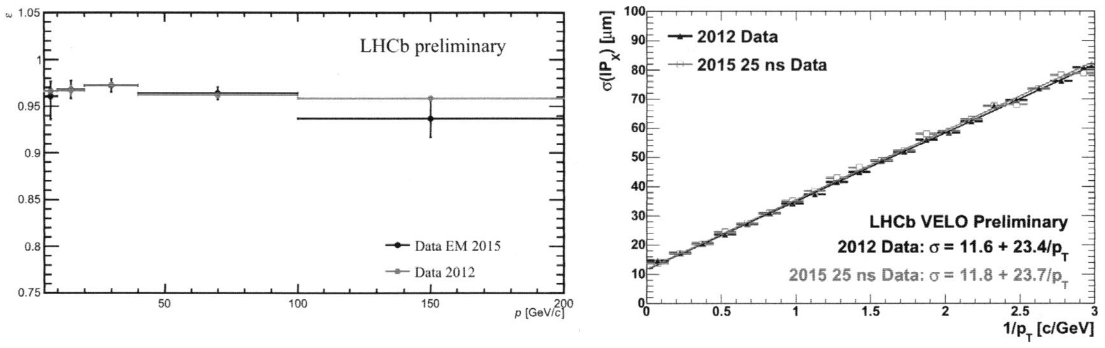

Figure 9. The performance of tracking system. Left: the 2015 tracking efficiency compared to 2012 data. Right: impact parameter resolution obtained for 2015 data with a comparison to 2012 data.

6 Turbo stream

The LHCb trigger for Run II contains a new streaming strategy (Turbo stream [6]) for many decays with a large production cross-section, which suffer due to limited trigger output rate.

In particular, in Run I the much smaller timing budget in the trigger system resulted in a low reconstruction efficiency of low-momentum charged particles, and the lack of full particle identification information in the HLT limited the performance for charm and strange hadron physics. Other issues were resolution differences between the online and the offline reconstruction, leading to difficulties in understanding efficiencies with a high degree of precision.

The new LHCb strategy resolves these problems and allows the selections to be performed with candidates reconstructed directly in the trigger, thus skipping the offline reconstruction. Improved reconstruction algorithms allowed to loosen the p_{T} threshold from $1.2\,\mathrm{GeV}/c$ to $500\,\mathrm{MeV}/c$ in HLT1, significantly improving the reconstruction efficiency of low-momentum particles. The high-statistic modes can now be selected exclusively using offline-quality selections. Only the candidates selected by the trigger are saved, thus reducing the amount of recorded data by a factor of 10 with no loss in physics performance. The selected candidates are available for data analysis 24 hours after data taking.

This new Turbo stream approach was successfully used in the first LHCb publications on 2015 data [7, 8]. In order to maximize the physics output of the experiment, it is planned to use a mixture of both inclusive and exclusive trigger selections in future, with particular use of the new streaming strategy.

7 Conclusion

To match the new running conditions, the strategy of data taking in the LHCb experiment has been redesigned during the long shutdown. New automatic real-time alignment and calibration procedures are used. This new scheme allows the events to be fully reconstructed with a quality equivalent to the Run I offline performance in the trigger, and enables signals to be selected with purities previously achieved only offline. Using the new Turbo stream and performing the full offline reconstruction in the trigger greatly increases the efficiency with which charm and strange hadron decays can be selected, thus broadening the physics programme of the LHCb experiment.

References

[1] A.A. Alves, Jr. et al. (LHCb collaboration), JINST **3**, S08005 (2008)
[2] R. Aaij et al. (LHCb collaboration), Int. J. Mod. Phys. **A30**, 1530022 (2015), arXiv:1412.6352
[3] R. Aaij et al., JINST **8**, P04022 (2013), arXiv:1211.3055
[4] J. Amoraal et al., Nucl. Inst. & Meth. A **712**, 48 (2013), arXiv:1207.4756
[5] W.D. Hulsbergen, Nucl. Inst. & Meth. A **600**, 471 (2009), arXiv:0810.2241
[6] R. Aaij et al., Comp. Phys. Comm. (2016), arXiv:1604.05596
[7] R. Aaij et al. (LHCb collaboration), JHEP **03**, 159 (2016), arXiv:1510.01707
[8] R. Aaij et al. (LHCb collaboration), JHEP **10**, 172 (2015), arXiv:1509.00771

EPJ Web of Conferences **127**, 00008 (2016) DOI: 10.1051/epjconf/201612700008
Connecting The Dots 2016

Associative Memory Pattern Matching for the L1 Track Trigger of CMS at the HL-LHC

Giacomo **Fedi**[1,a] on behalf of the CMS collaboration

[1] *INFN Sezione di Pisa, Italy*

Abstract. The High Luminosity LHC (HL-LHC) will deliver a luminosity of up to $5 \times 10^{34}\,\mathrm{cm}^{-2}\,\mathrm{s}^{-1}$, with an average of about 140 overlapping proton-proton collisions per bunch crossing. These extreme pileup conditions place stringent requirements on the trigger system to be able to cope with the resulting event rates. A key component of the CMS upgrade for HL-LHC is a track trigger system, able to identify tracks with transverse momenta above 2 GeV/c already at the first-level trigger. We present here the status of the implementation of a prototype system, based on the combination of Associative Memory custom ASIC and modern Field Programmable Gate Array (FPGA) devices, with the purpose to demonstrate the concept based on state-of-the-art technologies, and to direct the efforts of the necessary R&D toward a final system.

1 Introduction

The increase of the luminosity of the Large Hadron Collider (LHC) to $5 \times 10^{34}\,\mathrm{cm}^{-2}\,\mathrm{s}^{-1}$, foreseen in the High Luminosity (HL) upgrade scheduled for 2026, will bring up the number of proton-proton interactions per bunch crossing to an average of about 140 and a maximum of about 175. The corresponding higher background will generate an unmanageable increase in the trigger rate of the ATLAS and CMS detectors if their trigger systems will keep the current threshold configurations in order to preserve their physics-analysis capabilities.

One solution could come from the use of high-resolution spatial information from silicon trackers, at the cost of adding a few microseconds of latency and with a large data rate compared to the current L1-trigger. The trajectories of charged particles in CMS are helices since they are bent by the magnetic field generated by a solenoid. When a charged particle passes through the Silicon Tracker, it releases a small amount of energy in the silicon sensors, and the passage of the particle is recorded ("hit"). The CMS tracker for the HL-LHC upgrade will be composed of six (seven) layers of silicon modules in the central (forward) region. Each proton-proton interaction can generate hundreds of charged particles so that the Silicon Tracker generates thousands of hits in each event. Having a multitude of hits, the reconstruction of the trajectory of charged particles in few microseconds becomes challenging.

CMS is exploring three different approaches to tackle this problem: two are based on track reconstruction done with Field Programmable Gate Arrays (FPGAs), and one that will be explained in this report which is based on the use of Associative Memory (AM) and FPGAs. The purpose of our R&D activity is to prove that this goal can be achieved by combining pattern recognition based on an

[a] e-mail: giacomo.fedi@cern.ch

Figure 1. Schematic overview of part of the system. An ATCA shelf accommodates ten PulsarIIb boards and each of them can accommodate two PRMs. Two types of PRM have been developed: one with a full AM chip configuration and one with two Xilinx Ultascale FPGAs and a socket which can accommodate VIPRAM [5] chips.

AM custom chip and computing capabilities available in state-of-the-art FPGA devices. Track finding based on Associative Memories [1] has been successfully used in the CDF experiment [2] and, more recently, is being exploited in the Fast-Track processor [3, 4] for the ATLAS Level-2 trigger system.

The CMS outer tracker will be divided in 48 regions in η-ϕ (pseudo-rapidity and azimuthal angles) called trigger towers. In each trigger tower, the track finding will be performed using data from silicon modules belonging to that region. Each tower will receive an average of 100 hits per layer (the number of layers ranges from 6 to 7 depending on the pseudo-rapidity) in each bunch crossing. We want to build a prototype of the system with the available technology aiming to demonstrate that we will be able to build the full system for the HL phase of the LHC. Such a demonstration will make use of reasonable extrapolations of the technology progress and cost of commercial devices, and the assumption of a strong R&D of the AM chip.

2 Demonstrator system overview

Several boards based on the ATCA technology, the Pulsar IIb boards [6], will collect data from each trigger tower. Each Pulsar IIb board can accommodate two Pattern Recognition Mezzanine (PRM) boards performing pattern recognition and track fitting. Figure 1 shows a schematic view of the prototype system.

Each PRM board contains one or two Xilinx FPGAs, implementing data flow management and track fitting, and one or multiple AM devices. Each PRM board receives full resolution data and temporarily stores them in a data buffer. Lower resolution data, called "Super-Strips", are evaluated and transmitted to the AM chips. The AM chip behaves as a Content Addressable Memory (CAM) and quickly matches the Super-Strips with the pre-loaded patterns. Pre-loaded patterns are based on simulation of the tracks coming from the bunch crossing and of the tracker. Indeed, the tracks are mostly originating in a defined region of the detector, and the curvature of the helical tracks is confined to a certain region of interest ($p_{\mathrm{T}} > 2\,\mathrm{GeV}/c$).

All the patterns are simultaneously compared with the input data for each layer. When a match is found on a layer, a match bit is set and it remains as such until the end of the event processing, when an internal reset signal is propagated. If in a given pattern a match is detected for all the layers, or for a majority of them, the pattern address is transmitted to the FPGA.

Only a subset of the Super-Strips are associated to relevant tracks, mapped into patterns stored in the AM chips. The full resolution data associated to these Super-Strips are retrieved from the FPGA data buffer, filtered and propagated to the Track Fitting module. The processing time for each event in a PRM board will be of the order of few μs. Therefore it will be necessary to time-multiplex the data to O(10) Pulsar boards using a round-robin mechanism. Tracks that are found from the PRM will be then transferred to the Global Trigger System [7]. The final goal of the PRM board is to evaluate the performance of the real-time system described above using state-of-the-art AM chips and to shape the necessary modifications to match the tight CMS bandwidth requirement and processing time constraint of 4–5 μs.

For this purpose two prototype processors have been developed: the Pattern Recognition Mezzanine PRM05/06 at INFN and the FNAL PRM. The PRM05 processor is designed to combine the pattern recognition task [8] using the new version of the AM device with the precise track fitting [3], performed by the latest generation of FPGA devices, to send out the track information to the higher trigger levels within a latency of a few microseconds. PRM05 has a Xilinx Kintex-7 FPGA (XC7K355T) and 16 AM chips. The board has been developed to accommodate AM05 devices, to test the basic functionality of the serial links, even if the number of patterns stored in each chip is modest (2k patterns).

The current INFN PRM board houses 16 AM custom chips version 05 (AM05). Each chip can store up to 2k patterns, corresponding to a total capacity of 32k patterns per PRM (128k patterns in the AM version 06 chip, "AM06") to be released in the second quarter of 2016. Each pattern is made of 8 independent 16-bit words, one for each layer. The input data bus distribution from the FPGA to the AM chips uses a first stage of 1-to-2 fan-out buffers to distribute data to the left and right quarters of the card, followed by a second stage of two 1-to-4 fan-out buffers to distribute the data to the 4 AM chips in parallel. The input serial links can sustain a data rate of 2 Gbit/s, while the output data of each AM chip is connected directly to the FPGA using high speed serial lines running at 2.4 Gbit/s.

A second version of the INFN PRM will use a 12 AM chip version 06 where the number of patterns per chip will be increased to 128k patterns for a total of 1.5 million patterns per PRM, which will provide a high efficiency of track reconstruction. The FPGA has the role of distributing the hits to AM chips, collecting the candidate tracks, performing a track fitting on them using a Principal Component Analysis (PCA) [3] and then send out the results to the Global Trigger System. The direct connection between the FPGA and the AM chips has been chosen to eliminate daisy chains and to reduce as much as possible the latency of the Level-1 trigger decision and to integrate all the functionalities in one single board. An external low latency memory is added to increase the data storage capability of the device, in view of possible needs by the different fitting algorithms to be tested.

The FNAL PRM board has two Xilinx Ultrascale FPGAs and a socket which can accommodate the next-to-come VIPRAM [5] chip. One of the FPGA is used to emulate the AM chip using synthesized HDL code, while the other FPGA is used to store the hits and to fit the matched patterns. The emulated AM chip can store up to 4k patterns. This mezzanine cannot manage the patterns of a full trigger tower with high efficiency, but is designed to minimize the latency of the pattern finding process. Indeed, the FNAL PRM is equipped with a powerful FPGA with high speed serial links with data rates up to 16.3 Gbit/s, and the synthesized FPGA code has been designed with a low latency approach. With this PRM we want to show what we can achieve with the future AM custom chips.

DOI: 10.1051/epjconf/201612700008

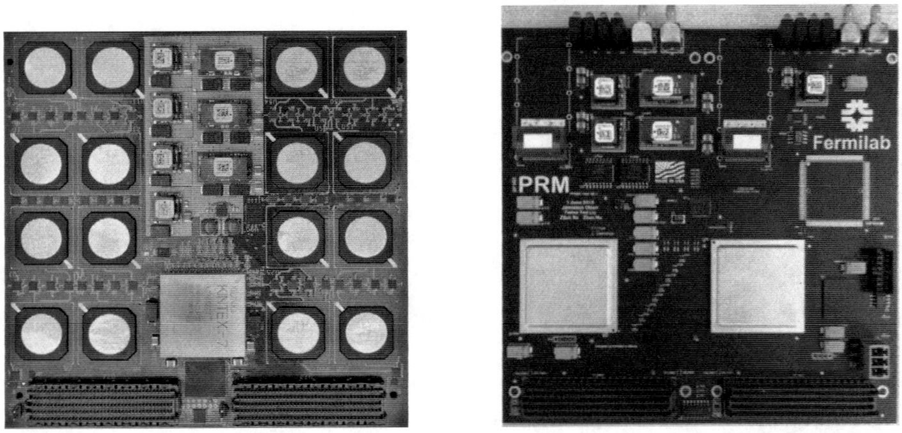

Figure 2. Left: INFN pattern recognition mezzanine board. Right: FNAL pattern recognition mezzanine board.

3 PRM architecture

The PRM board, see Figure 2, is a custom $14.9 \times 14.9\,\text{cm}^2$ printed circuit board housing the following components:

- FMC connectors (2x);
- Xilinx FPGA (1x INFN PRM, 2x FNAL PRM);
- Associative Memory devices (16x INFN PRM, 1x FNAL PRM);
- Power regulators;
- DDRII Pattern Memory (1x);

The two FMC connectors, compliant with the VITA 57.1 standard, are used to interface the PRM to the Pulsar IIb host board. They carry signals and power supply voltages. The host board provides 3.3 V and 12 V, for a maximum available power of 150 W. The power regulator group is used to generate the 1.0 V, 1.2 V, 1.8 V and 2.5 V voltages required by the FPGA, the AM devices and the pattern memory. Regarding the signal lines, 6 bidirectional high-speed serial links (up to 10 Gbit/s), with three links through each FMC connector, are used to send and retrieve data from the PRM with a total bandwidth in input and output up to 120 Gbit/s. Moreover, 68 additional LVDS pair pins are used for the slow configuration and monitoring purposes of the PRM and to provide additional bandwidth for the data transfer. The Xilinx FPGA manages the data flow inside the mezzanine. It receives and temporarily stores the full resolution hit data from the Pulsar IIb board and it evaluates and distributes to the 16 AM chips the lower resolution hit data used in the pattern recognition. Finally, the external pattern memory provides up to 18 Mbit for INFN PRM and up to 36 Mbit for FNAL PRM of memory resources to store a copy of the pattern banks of the AM chips.

The FNAL RPM is equipped with two Xilinx Ultrascale 040/060 FPGAs, fast serial links connect the two FPGA with a data rate up to 16.3 Gbit/s. One of the FPGA is also fully connected with six high-speed serial links to the FMC connectors. The PRM is also provided with additional external communication links based on QSFP+ transceivers.

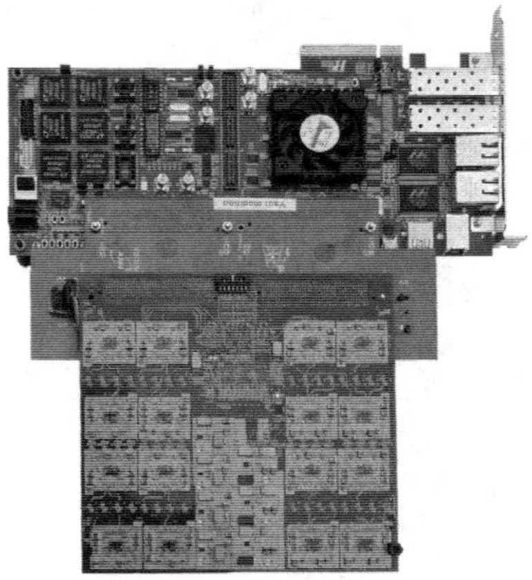

Figure 3. PRM test stand. Top: the HTG Virtex-6 evaluation board; middle: the adapter card, bottom: the INFN PRM.

4 PRM hardware validation

Both INFN and FNAL PRM boards were validated regarding the electrical circuitry and both showed the expected behavior. The JTAG connection to the FPGA was tested by programming it.

The INFN PRM board was tested using the test stand shown in Figure 3. The test stand is composed of a PRM board, an adapter card, and an evaluation board. An HTG Virtex-6 evaluation board has been used for the main tests reported in this text. Since the HTG evaluation board has some limitation on the links towards the FMC connectors (only four high speed serial links connected and data rate limited to 6.2 Gbit/s), in order to test the fast serial links of the INFN PRM, an Ultrascale Evaluation board has been also used. The evaluation board is connected to the host PC via Ethernet. It provides power, LVDS parallel connections and high-speed serial connections to the PRM board. The IP-bus protocol is used to access devices in the Virtex-6 and in the PRM FPGAs (e.g.: control and status registers, source FIFOs containing test hits, monitoring FIFOs containing PRM outputs). All the serial links between the FPGAs and to/from the AM chips on the PRM board have been successfully tested. JTAG connections to the AM chips (4 JTAG chains with 4 AM chips each) have been tested in order to verify the possibility to configure and program the AM chips. The eight input links to each AM chip have been characterized. Pseudo Random Bit Sequences (PRBS) have been used to test the links between the FPGA and the AM chips. The FMC connector links have undergone tests by using a loopback card and the IBERT tool, provided by Xilinx, that allows measuring the Bit Error Ratio (BER) and producing the corresponding eye diagrams. Signal integrity up to 8 Gbit/s has been tested by measuring the BER and the corresponding eye-scans in the receiver links of the PRM and in the receiver links of the Ultrascale board. The BER has been measured using a PRBS-7 sequence. The measured BER was less than 2×10^{-15} in both directions. The power consumption of the board has been measured by powering and configuring the four groups of the AM chips, resulting in a static power consumption of about 40 W. The mezzanine board has been successfully integrated with the Pulsar IIb board in Summer 2015. During these tests the mechanical and electrical compatibility have been verified and both the LVDS and high-speed differential pairs have been fully tested.

The FNAL PRM board has been tested in a standalone configuration and with the Pulsar IIb board. The high speed serial links between the two FPGAs and towards the FMC connectors have been tested: between the FPGAs the links were validated at 16.3 Gbit/s while towards the FMC connectors they were validated at 10 Gbit/s. The LVDS links between the FPGAs and the FMC connectors were successfully validated. The QSFP+ transceivers were also tested and no errors were found in the transmitted pseudo-data.

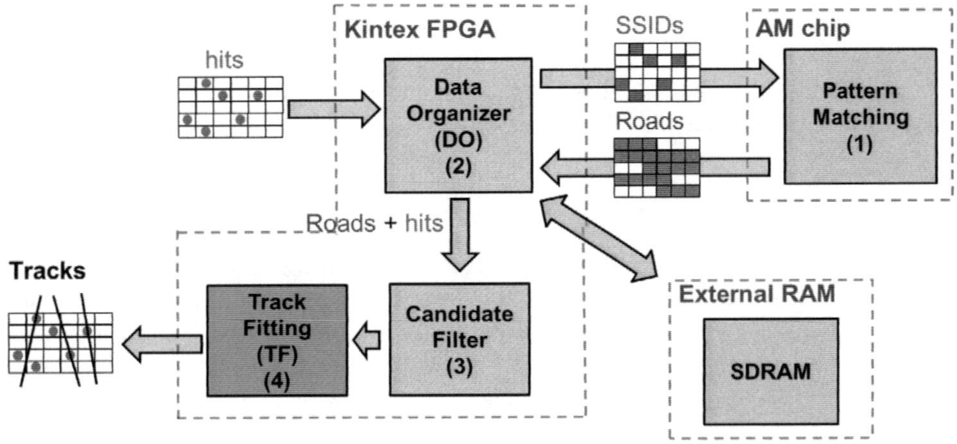

Figure 4. Schematic view of the modules of the track finding algorithm.

5 PRM functionalities

On both INFN and FNAL PRM boards a synthesized HDL code has been developed to implement functionalities needed for the track finding: hit memorization, hit retrieving, and hit fitting. The tracking algorithm implemented in both PRM boards consists of the following sequence of operations, see Figure 4:

1. Full resolution hit data received by the host board are decoded and stored in a smart data buffer, called Data Organizer, and coarse resolution hit data, "Super-Strips", are generated and transmitted to the AM chips;

2. AM chips perform pattern recognition on Super-Strip input data and identify sets of Super-Strips matching patterns, called "Road", that are transmitted back to the FPGA;

3. Roads are used to retrieve the associated Super-Strips from an external pattern memory (currently internal to the FPGA) the associated Super-Strips;

4. Hit data stored in the Data Organizer and belonging to the Super-Strips retrieved from the pattern memory are propagated to the filtering module;

5. The Super-Strips belonging to a road are filtered in such a way that only one candidate track is generated composed of no more than one hit per layer. The set of hits belonging to the candidate track is sent to the track fitter module;

6. The track fitter performs the fitting algorithm using the full resolution data from the Data Organizer to evaluate the track helix parameters and the goodness of the fit.

Two different track finding algorithm were developed. Each one was dedicated to fulfill the goal of the PRM boards. In the case of the FNAL PRM board the algorithm focused on the reduction of the track finding latency, while in the case of INFN PRM board the algorithm focused on the handling of large pattern banks and multiple AM devices. The full chain of track finding has been developed and tested using HDL simulations and the real FPGAs for simple events. In both cases the latency owing to the hit retrieving, hit filtering, and hit fitting was about one microsecond. In the case of the INFN mezzanine additional and non-negligible latency is caused by the limited bandwidth of the serial links and a non-optimized output electronics of the AM05 device. It is worth mentioning that the AM05 chip has been developed for applications where the latency is not the main concern. An R&D phase to develop a faster and low-latency AM device is already ongoing.

6 Conclusions

The INFN and FNAL PRM boards and the related software have been developed and tested. Further optimization of the algorithms and a new INFN PRM board equipped with AM06 will be done leading to the system prototype to be developed before the end of this year (2016). The current hardware and algorithm showed good performance and limited latency, demonstrating promising results in view of the final test of the system prototype.

Acknowledgments

This work is partially funded by MIUR Italy, under contract 2012Z23ERZ of PRIN 2012 H-TEAM and the People Program (Marie Curie Actions) of the European Union's Seventh Framework Program FP7/2007-2013/ under REA grant agreement no. 317446 INFIERI "INtelligent Fast Interconnected and Efficient Devices for Frontier Exploitation in Research and Industry".

References

[1] M. Dell'Orso, L. Ristori, Nucl. Inst. & Meth. A **278**, 436 (1989)
[2] J.A. Adelman et al. (CDF collaboration), Nucl. Inst. & Meth. A **572**, 361 (2007)
[3] C. Gentsos, F. Crescioli, P. Giannetti, D. Magalotti, S. Nikolaidis, PoS **TIPP2014**, 209 (2014)
[4] F. Alberti et al., JINST **8**, C01040 (2013)
[5] T. Liu, G. Deptuch, J. Hoff, S. Jindariani, S. Joshi, J. Olsen, N. Tran, M. Trimpl, JINST **10**, C02029 (2015)
[6] J. Olsen, T. Liu, Y. Okumura, JINST **9**, C01041 (2014)
[7] M. Jeitler et al., JINST **2**, P01006 (2007)
[8] A. Andreani et al., JINST **9**, C11006 (2014)

EPJ Web of Conferences **127**, 00009 (2016) DOI: 10.1051/epjconf/201612700009
Connecting The Dots 2016

Boosted Jet Tagging with Jet-Images and Deep Neural Networks

Michael **Kagan**[1,a], Luke **de Oliveira**[2], Lester **Mackey**[2], Benjamin **Nachman**[1,2,b], and Ariel **Schwartzman**[1,c]

[1] *SLAC National Accelerator Laboratory, Menlo Park, CA, USA*
[2] *Stanford University, Stanford, CA, USA*

Abstract. Building on the jet-image based representation of high energy jets, we develop computer vision based techniques for jet tagging through the use of deep neural networks. Jet-images enabled the connection between jet substructure and tagging with the fields of computer vision and image processing. We show how applying such techniques using deep neural networks can improve the performance to identify highly boosted W bosons with respect to state-of-the-art substructure methods. In addition, we explore new ways to extract and visualize the discriminating features of different classes of jets, adding a new capability to understand the physics within jets and to design more powerful jet tagging methods.

1 Introduction

Jets, or collimated streams of particles arising from the production of high energy quarks and gluons, are crucial signatures for discovering signs of new physics at the Large Hadron Collider (LHC) [1]. Quarks and gluons, and subsequently jets, are produced ubiquitously from the LHC proton-proton collisions. However, in many theories of new physics, new TeV-mass-scale particles decay into high energy $W^{\pm}$, Z, H bosons, and top quarks. These high energy bosons and top quarks can subsequently decay into single jets containing all of their decay products. Identifying, or *tagging*, the jets from bosons and top quarks is thus vital to reconstruct and identify signs of new TeV-mass-scale particles. The field of jet-substructure (see [2–4] and references therein) has emerged to overcome this challenge, whereby physics inspired features are engineered to exploit the differences in the internal structure of jets produced by bosons and top quarks from that of jets produced directly from quarks and gluons. Concurrently, modern computer vision and machine learning techniques have made great progress in recent years in deep learning [5–7]; large multi-hidden layer neural networks operating on low level information (e.g. pixels in an image), without feature engineering, have been found to learn complex representations of information within images and efficiently use this information for classification.

The jet-images approach to jet tagging combines computer vision with jet-substructure. Jet measurements from LHC detectors are formatted into images, where detector energy measurements play

[a] e-mail: makagan@slac.stanford.edu
[b] e-mail: bpn7@slac.stanford.edu
[c] e-mail: sch@slac.stanford.edu

the role of pixel intensities, and the jet-images are analyzed using computer vision inspired techniques. The computer vision techniques include linear methods, such as Fisher discriminant analysis, and deep learning methods, including convolutional and dense MaxOut neural network architectures. These techniques not only provide improved discrimination power, but also insight into the physics learned for discrimination.

The power of the jet-images and computer vision techniques is explored in a benchmark jet-substructure problem, discriminating high energy $W^{\pm}$ bosons from quarks and gluons, as discussed in references [8] and [9]. We refer the reader to these references for further details.

2 Samples and selection

To study the jet-images technique, the benchmark task of discriminating high energy $W^{\pm}$ bosons from quarks and gluons is examined. Monte Carlo (MC) simulations of proton-proton collisions at $\sqrt{s} = 14\,\text{TeV}$ based on the Pythia 8.170 [10, 11] event generator are used. High energy $W^{\pm}$ bosons are simulated through the decay of a hypothetical new heavy W' boson which decays to one $W^{\pm}$ boson and one Z boson. Subsequently, the W is forced to decay hadronically ($W \to qq'$), and the Z to decay invisibly ($Z \to \nu\nu$). The background is the generic (multijet) production of quarks and gluons.

Since the classification performance can depend on the transverse momentum, p_{T}, and jet mass, m, of the jets, we focus our studies on a restricted range of $250\,\text{GeV}/c < p_{\mathrm{T}} < 300\,\text{GeV}/c$, and a $65\,\text{GeV}/c^2 < m < 95\,\text{GeV}/c^2$ mass window that contains the peak of the W. The minimum angular spread of the W decay products is approximately $\Delta R = \sqrt{\Delta\eta^2 + \Delta\phi^2} \approx 0.6$, where $\Delta\eta$ and $\Delta\phi$ are the distances between W boson decay products in (η, ϕ) coordinates. Further, to remove any dependence on the jet transverse momentum, we re-weight the signal to have the same p_{T} distribution as the background.

The discrete size and finite acceptance of a real detector are modeled by binning the particles into cells with size 0.1×0.1 in (η, ϕ) extending out to $\eta = 5.0$. Each cell's 4-vector is defined by $p_j = \sum_{i \text{ incident on } j} E_i(\cos\phi_j/\cosh\eta_j, \sin\phi_j/\cosh\eta_j, \sinh\eta_j/\cosh\eta_j, 1)$, where the sum runs over particles indent on the cell.

Jets are built by clustering the cells using the anti-k_t algorithm [12] with distance parameter $R = 1.0$, via FastJet [13] 3.0.3. The large distance parameter is used in order to capture all the decay products of a boosted W boson. To reduce the impact of pileup[1] and underlying event, jets are are trimmed [14] by re-clustering the constituents into $R = 0.3$ k_t subjets and dropping those which have $p_{\mathrm{T}}^{\text{subjet}} < 0.05 \times p_{\mathrm{T}}^{\text{jet}}$.

In order to compare the jet-images approach with standard substructure techniques, we compare performance with known highly performant jet substructure features. These features are the *jet mass* defined as $m_{\text{jet}}^2 = \sum_{i,j} p_i p_j$ with jet constituent four-vectors p_i, *n-subjettiness* [15] in the form of τ_{21} using the winner-take-all axis definition [16], and the distance in (η, ϕ) space between subjets of the trimmed jet (ΔR).

3 Jet-images and pre-processing

Jet-images are formed by taking a 25×25 cell grid around the jet axis, with the intensity of jet-image pixel i defined to be the corresponding cell transverse energy $p_{\mathrm{T},i} = E_i/\cosh(\eta_i)$.

In order for machine learning algorithms to learn more efficiently, we make use of space-time symmetries to pre-process the jet-images into a standardized representation. The pre-processing steps are:

[1] Pileup is generated by multiple proton-proton collisions per bunch crossing

- *Translation*: Jet-images are translated such that the leading p_T subjet is located at $(\eta, \phi) = (0, 0)$. Note that translations in ϕ are effectively rotations around the detector z-axis, while translations in η are Lorentz boosts along the z-axis. As such, translations in η can alter the mass of a jet-image if the pixel energies are kept fixed. However, the transverse energy is invariant to such η translations, and thus is used for the pixel intensities.
- *Rotation*: Jets are rotated such that the second leading p_T subjet is aligned along the vertical axis of the jet-image. If the jet has only one subjet, the first principle component of the energy distribution in (η, ϕ) is rotated to align with the vertical axis of the jet-image.
- *Parity Flip*: After rotation, images are flipped over the vertical axis such that the right side of the image has a higher energy than the left. This helps to standardize the location of additional radiation in the jet-image.

After pre-processing, the leading subjet within the jet-image is located at the center of the image, and the second subjet (if it exists) is aligned along the vertical axis of the image. In facial recognition tasks, this is equivalent to aligning the eyes within an image of a face. With such a standardized jet-image representation, the machine learning algorithms do not need to learn about the symmetries in the jet-image, thus allowing the learning to focus more effectively on discrimination. The effect of the pre-processing on a collection of W jets can be found in Figure 1.

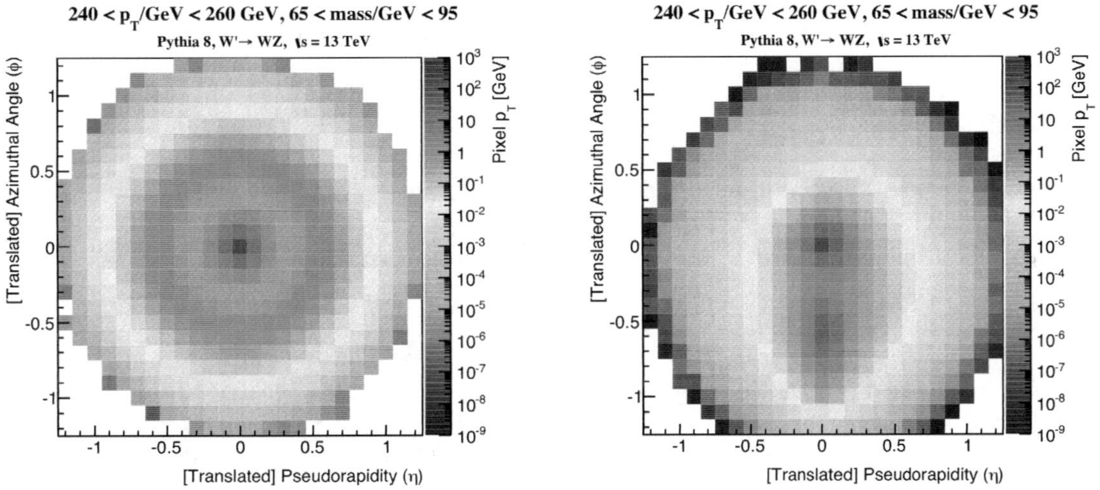

Figure 1. The average jet-image for W jets before (left) and after (right) pre-processing. The average is taken over jet-images with $240\,\text{GeV}/c < p_T < 260\,\text{GeV}/c$ and $65\,\text{GeV}/c^2 < m < 95\,\text{GeV}/c^2$.

4 Deep learning architectures and training

Discrimination between W and quark/gluon jet-images is performed using deep neural networks (DNN), which have been found to outperform competing algorithms in computer vision tasks similar to jet tagging with jet-images. DNNs have been found to learn rich high-level representations from raw (pixel-level) image data [5–7]. We make use of the power of such networks by training them on jet-images, with the pixel level information as input. We focus our attention here on two state-of-the-art architectures: fully connected MaxOut networks [6], and convolutional neural networks [5].

Complete descriptions of the architectures can be found in reference [9], and are only briefly described here. The MaxOut network consists of two MaxOut layers followed by two fully connected layers, followed by a classification layer. The convolutional network (ConvNet) consists of three convolution layers, followed by one fully connected layer, followed by a classification layer. Convolution kernel sizes were 11×11 in the first convolutional layer, and 3 in subsequent convolutions. The atypically large kernel size in the first convolution was found to help performance on the relatively sparse jet-images (only about 5–10% of pixels contain energy on average in a jet-image). Each convolution is followed by a max-pooling layer [17] and then by a dropout layer [7]. All MaxOut, dense, and convolutional layers use the rectified linear unit (ReLU) activation function [18], while the classification layer uses sigmoid activations.

All deep learning experiments were conducted in Python with the Keras [19] deep learning library, utilizing NVIDIA C2070 graphics cards. 8 million examples were used for training, with an additional 2 million validation samples for tuning the hyper-parameters, and 3 million examples were used for testing. Signal examples were weighted such that the total sum of weights was the same as the total number of background examples. These weights were used by the cost function in the training and in the ROC curve computations of the test samples. The networks were trained with the Adam [20] algorithm. The training consisted of 100 epochs, with a 10 epoch patience parameter on the increase in area under the ROC curve between 0.2 and 0.8 on a validation set. Batch sizes of 32 were used for the MaxOut network, while batch sizes of 96 were used for the convolution networks.

5 Results

The primary figure of merit used in this note to compare the performance of different classifiers is the ROC curve, which shows the quark/gluon background rejection (the inverse misclassification efficiency) as a function of the W jet signal efficiency. Efficiencies and rejection are computed by scanning thresholds on the signal-to-background likelihood ratio distribution (i.e. the optimal use of the variables).

The ROC curve showing the DNNs as well as several substructure variables can be found on the left in Figure 2. Both DNNs are seen to have significantly better rejection than the substructure variables for all efficiencies, with an improvement in rejection of about 2 at a signal efficiency of 30%. We also see that the MaxOut network tends to outperform the ConvNet, which we believe is due to difficulties for convolutional networks to learn rich discriminating information from kernels when images are sparse.

In order to understand if the DNNs have learned the information contained within the substructure variables, we combine the DNN outputs with substructure variables into a single discriminant using the likelihood ratio of the 2D distribution of the DNN and the substructure variable. The combination with the ConvNet is shown on the right in Figure 2. We can see that the combination with ΔR and τ_{21} adds little discrimination to the baseline performance of the ConvNet, indicating that the ConvNet has learned the discriminating information contained in these variables. However, when the jet mass is combined with the ConvNet, the performance is seen to increase, indicating that the ConvNet has not completely learned the mass information. Similar behavior is found for the MaxOut network.

To examine more deeply into the information learned by the DNNs, Figure 3 shows the 2D distribution of τ_{21} (left) and jet mass (right) conditioned on the ConvNet output, i.e. $p(X|\text{DNN output})$ where X is the substructure variable. As we can see, the ConvNet output is strongly correlated with τ_{21}, again indicating that this information has been learned by the network. However, the correlation

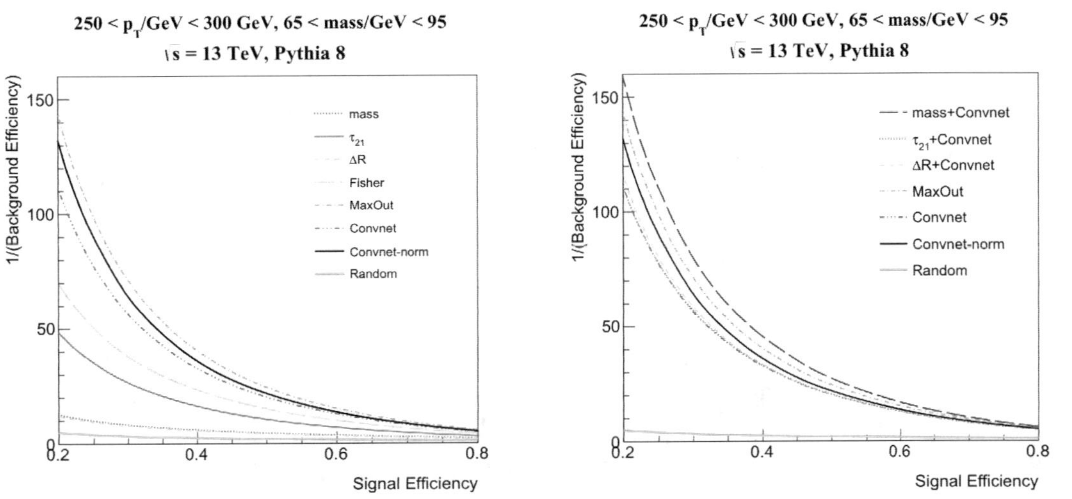

Figure 2. ROC curves comparing DNNs with substructure discriminants (left), and ROC curves when DNNs are combined with substructure variables (right).

with mass is not as strong, with the distributions of mass significantly spread out over various ConvNet output values. While some correlation is present, the network does not seem to have learned all information contained in the jet mass.

In order to study what is learned by the DNNs beyond known substructure variables, we examine the networks in several ways. First, we find the 500 most activating images for a given neuron,

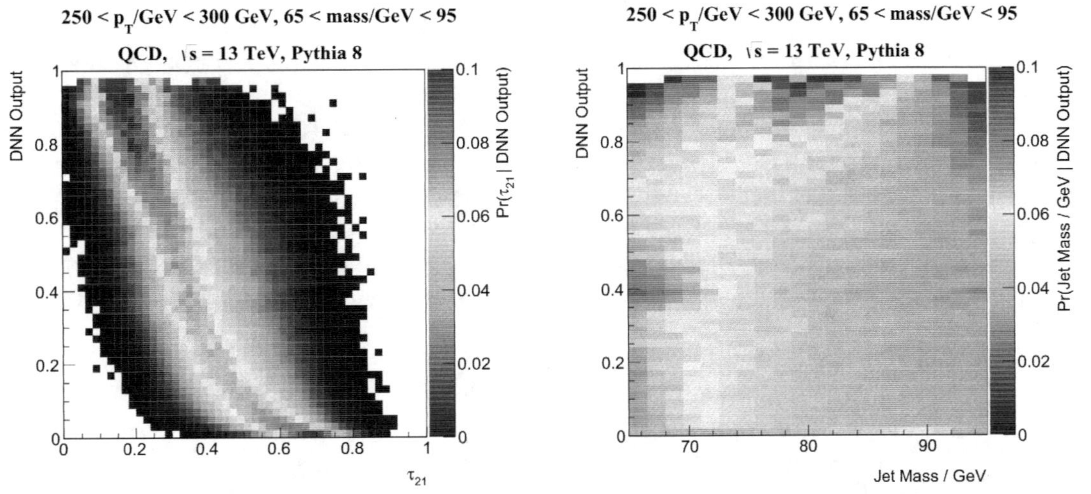

Figure 3. 2D distribution of τ_{21} (left) and jet mass (right) conditioned on the ConvNet output, i.e. $p(X|\text{DNN output})$ where X is the substructure variable)

and visualize the average of these images. This is shown in Figure 4 for several neurons in the last hidden layer of the ConvNet. Also shown above each images is the fraction of images found to be most activating for each neuron that were signal images. As we can see, the signal correlated neurons have a clear two-prong structure with information between the leading two subjets but very little information outside of the two-prong structure. In contrast, background correlated neurons have much broader energy distributions indicating the presence of additional wide-angle radiation beyond the leading two subjets.

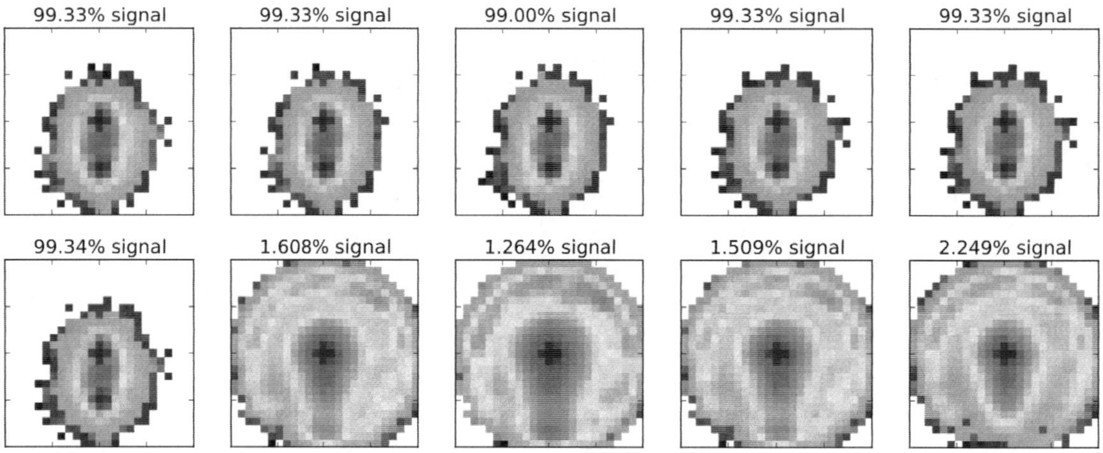

Figure 4. Average most activating input images for several neurons in the last hidden layer of the ConvNet.

To further understand how the DNNs use the information present in the jet-images for discrimination, we compute the Pearson correlation coefficient for each jet-image pixel with the output of the DNN. These correlations with the ConvNet can then be visualized as an image, as seen in Figure 5. From this correlation image, we see that the presence of a leading subjet at the center of the image provides little discrimination power. This is not surprising, as both signal and background jets have at least one high energy subjet. However, the location of the subleading subjet is strongly correlated with signal-like images, as seen by the red region at $(\eta, \phi) \approx (-0.5, 0)$. In addition, radiation present around the the leading subjet and radiation at the periphery of the image away from either subjet is correlated with background like images. This indicates that the radiation pattern beyond the leading subjets provides important information for discrimination.

While we have focused on jets with $240\,\mathrm{GeV}/c < p_\mathrm{T} < 260\,\mathrm{GeV}/c$ and $65\,\mathrm{GeV}/c^2 < m < 95\,\mathrm{GeV}/c^2$, we can restrict the phase space of the images in such a way as to remove discriminating information from known powerful features. In doing so, we can examine the information learned by the DNNs beyond the known substructure features. As such we restrict the phase space to a signal-like region of $240\,\mathrm{GeV}/c < p_\mathrm{T} < 260\,\mathrm{GeV}/c$, $79\,\mathrm{GeV}/c^2 < m < 81\,\mathrm{GeV}/c^2$, and $0.19 < \tau_{21} < 0.21$. The ROC curve comparing classification performance, as well as the DNN-pixel correlation image, can be found in Figure 6. In this phase space, we can see that mass and τ_{21} have little discrimination power, as expected in such a restricted region. However, we can also see that the DNNs still provide discrimination power, further indicating that the DNNs have learned physics information beyond these known substructure variables. The ConvNet-pixel correlation image can help to elucidate what information has been learned. In this phase space, we see again that the radiation pattern provides important discrimination power. The radiation between subjets is correlated with signal-like images,

DOI: 10.1051/epjconf/201612700009

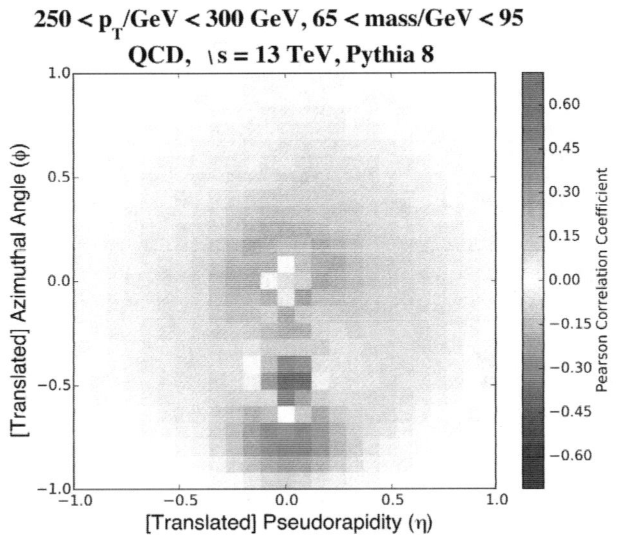

Figure 5. Pearson correlation coefficient of each input pixel with the ConvNet output.

while radiation at the periphery is correlated with background like images. This could be due to the difference between radiation patterns of the decays of a color singlet *W* boson, as apposed to the decays of a color octet gluon or color triplet quark. While features engineered to exploit such colorflow information have been studied in the past [21], the features learned here may present a new way to access this information for improved discrimination.

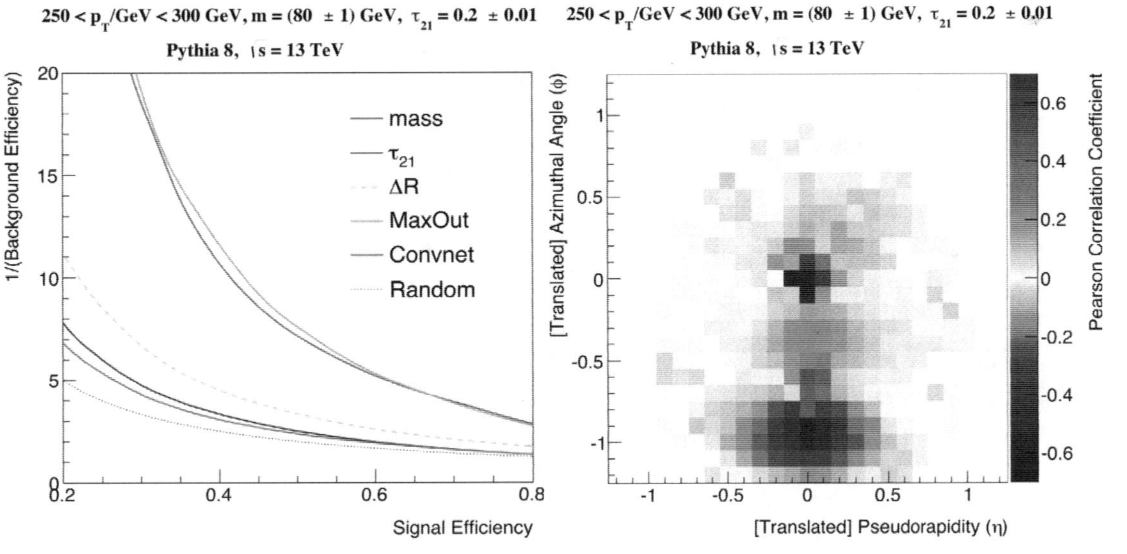

Figure 6. ROC curve comparing DNN and substructure variable performance (left), and Pearson correlation coefficient of each input pixel with the ConvNet output (right). Both figures are computed in the restricted phase space of $240\,\mathrm{GeV}/c < p_{\mathrm{T}} < 260\,\mathrm{GeV}/c$, $79\,\mathrm{GeV}/c^2 < m < 81\,\mathrm{GeV}/c^2$, and $0.19 < \tau_{21} < 0.21$.

6 Conclusions

In this paper, we have used the paradigm of jet-images to interpret LHC data based on tools inspired by image processing, computer vision, and deep learning. Deep learning proved to be powerful paradigm for jet tagging when combined with jet-images, significantly improving performance over state-of-the-art jet substructure features. The performance of deep learning techniques was aided by domain specific inputs, whereby physics inspired pre-processing allowed deep neural networks to learn powerful discriminating information without having to simultaneously learn the symmetries of space-time. Finally, we showed several techniques to extract information about what the deep neural networks are learning. A key aspect to this was being able to visualize the learned information in several ways.

Acknowledgments

This work was supported by the Stanford Data Science Initiative and by the US Department of Energy (DOE) grant DE-AC02-76SF00515.

References

[1] L. Evans, P. Bryant, JINST **3**, S08001 (2008)
[2] A. Altheimer et al., J. Phys. **G39**, 063001 (2012), arXiv:1201.0008
[3] A. Altheimer et al., Eur. Phys. J. **C74**, 2792 (2014), arXiv:1311.2708
[4] D. Adams et al., Eur. Phys. J. **C75**, 409 (2015), arXiv:1504.00679
[5] K. Simonyan, A. Zisserman (2014), arXiv:1409.1556
[6] I.J. Goodfellow, D. Warde-Farley, M. Mirza, A. Courville, Y. Bengio (2013), arXiv:1302.4389
[7] G.E. Hinton, N. Srivastava, A. Krizhevsky, I. Sutskever, R. Salakhutdinov (2012), arXiv:1207.0580
[8] J. Cogan, M. Kagan, E. Strauss, A. Schwarztman, JHEP **02**, 118 (2015), arXiv:1407.5675
[9] L. de Oliveira, M. Kagan, L. Mackey, B. Nachman, A. Schwartzman (2015), arXiv:1511.05190
[10] T. Sjostrand, S. Mrenna, P.Z. Skands, Comput. Phys. Commun. **178**, 852 (2008), arXiv:0710.3820
[11] T. Sjostrand, S. Mrenna, P.Z. Skands, JHEP **0605**, 026 (2006), arXiv:0603175
[12] M. Cacciari, G.P. Salam, G. Soyez, JHEP **0804**, 063 (2008), arXiv:0802.1189
[13] M. Cacciari, G.P. Salam, G. Soyez, Eur. Phys. J. **C72**, 1896 (2012), arXiv:1111.6097
[14] D. Krohn, J. Thaler, L.T. Wang, JHEP **1002**, 084 (2010), arXiv:0912.1342
[15] J. Thaler, K. Van Tilburg, JHEP **1103**, 015 (2011), arXiv:1011.2268
[16] A.J. Larkoski, D. Neill, J. Thaler, JHEP **04**, 017 (2014), arXiv:1401.2158
[17] D. Scherer, A. Müller, S. Behnke, *Evaluation of pooling operations in convolutional architectures for object recognition*, in *Artificial Neural Networks - ICANN 2010* (Springer, New York, 2010)
[18] X. Glorot, A. Bordes, Y. Bengio, Journal of Machine Learning Research **15**, 315 (2011)
[19] F. Chollet, *Keras*, https://github.com/fchollet/keras (2015)
[20] D.P. Kingma, J. Ba, Proceedings of the 3rd International Conference for Learning Representations (2014), arXiv:11412.6980
[21] J. Gallicchio, M.D. Schwartz, Phys. Rev. Lett. **105**, 022001 (2010), arXiv:1001.5027

Kalman Filter Tracking on Parallel Architectures

Giuseppe **Cerati**[1,a], Peter **Elmer**[2,b], Slava **Krutelyov**[1,c], Steven **Lantz**[3,d], Matthieu **Lefebvre**[2,e], Kevin **McDermott**[3,f], Daniel **Riley**[3,g], Matevž **Tadel**[1,h], Peter **Wittich**[3,i], Frank **Würthwein**[1,j], and Avi **Yagil**[1,k]

[1]*UC San Diego, La Jolla, CA, USA*
[2]*Princeton University, Princeton, NJ, USA*
[3]*Cornell University, Ithaca, NY, USA*

Abstract. Power density constraints are limiting the performance improvements of modern CPUs. To address this we have seen the introduction of lower-power, multi-core processors such as GPGPU, ARM and Intel MIC. In order to achieve the theoretical performance gains of these processors, it will be necessary to parallelize algorithms to exploit larger numbers of lightweight cores and specialized functions like large vector units. Track finding and fitting is one of the most computationally challenging problems for event reconstruction in particle physics. At the High-Luminosity Large Hadron Collider (HL-LHC), for example, this will be by far the dominant problem. The need for greater parallelism has driven investigations of very different track finding techniques such as Cellular Automata or Hough Transforms. The most common track finding techniques in use today, however, are those based on a Kalman filter approach. Significant experience has been accumulated with these techniques on real tracking detector systems, both in the trigger and offline. They are known to provide high physics performance, are robust, and are in use today at the LHC. Given the utility of the Kalman filter in track finding, we have begun to port these algorithms to parallel architectures, namely Intel Xeon and Xeon Phi. We report here on our progress towards an end-to-end track reconstruction algorithm fully exploiting vectorization and parallelization techniques in a simplified experimental environment.

[a]e-mail: giuseppe.cerati@cern.ch
[b]e-mail: peter.elmer@cern.ch
[c]e-mail: vyacheslav.krutelyov@cern.ch
[d]e-mail: steve.lantz@cornell.edu
[e]e-mail: ml15@princeton.edu
[f]e-mail: kevin.mcdermott@cern.ch
[g]e-mail: daniel.riley@cornell.edu
[h]e-mail: matevz.tadel@cern.ch
[i]e-mail: wittich@cornell.edu
[j]e-mail: fkw@ucsd.edu
[k]e-mail: ayagil@physics.ucsd.edu

1 Introduction

The Large Hadron Collider (LHC) at CERN is the highest energy collider ever constructed. It consists of two counter-circulating proton beams made to collide in four locations around a 27 kilometer ring straddling the border between Switzerland and France. By several means it is the largest man-made scientific device on the planet. The goal of the LHC is to probe the basic building blocks of matter and their interactions. In 2012, the Higgs boson was discovered by the CMS and ATLAS collaborations [1, 2]. By measuring the energy and momentum of particles escaping the collision, we can infer the existence of massive particles that were created and measure those massive particles' properties. This process is known as event reconstruction and consists of integrating information from different detector components. Track reconstruction, also known as tracking, is one step in event reconstruction and determines the trajectories of charged particles ("tracks") from a set of positions of energy deposits within the various layers of our detector ("hits"). Tracking, in comparison to other aspects of event reconstruction, is the most computationally time consuming, the most sensitive to increased activity in the detector, and traditionally, the least amenable to parallelized processing. The speed of online reconstruction has a direct impact on how much data can be stored from the 40 MHz collisions rate, while the speed on the offline reconstruction limits how much data can be processed for physics analyses. This research is aimed at vastly speeding up tracking.

The large time spent in tracking will become even more important in the HL-LHC era of the Large Hadron Collider. The increase in event rate will lead to an increase in detector occupancy ("pile-up", PU), leading to an exponential gain in time taken to perform track reconstruction, as can be seen in Figure 1 [3]. In the Figure, PU25 corresponds to the data taken during 2012, and PU140 corresponds to the low end of estimates for the HL-LHC era. Clearly this research on tracking performance will become increasingly important during this era.

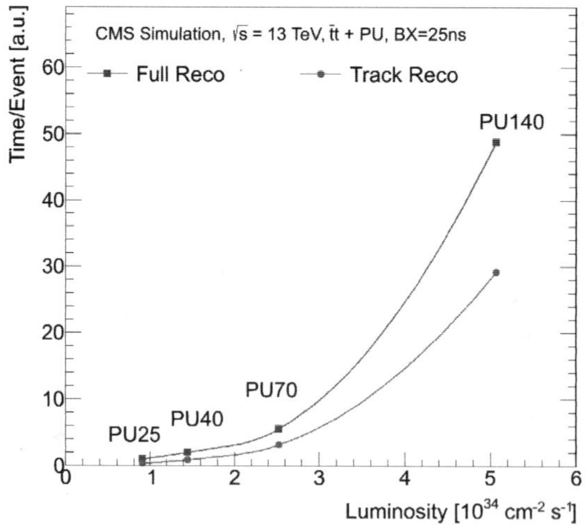

Figure 1. CPU time per event versus instantaneous luminosity, for both full reconstruction and the dominant tracking portion. PU25 corresponds to the data taken during 2012, and PU140 corresponds to the HL-LHC era. The time of the reconstruction is dominated by track reconstruction [3].

A correlated issue to speeding up tracking is the change in computing architectures in the last decade. Around 2005, the computing processor market reached a turning point: power density limitations in chips ended the long-time trend of ever-increasing clock speeds, and our applications no longer immediately run exponentially faster on subsequent generations of processors. This is true even though the underlying transistor count continues to increase per Moore's law. Increases in processing speed such as those required by the time increases in Figure 1 will no longer come 'for free' from faster processors. New, parallel processors instead are aggregates of 'small cores' that in total still show large performance gains from generation to generation, but their usage requires a re-work of our mostly serial software to exploit these gains. The processors in question include ARM, GPGPU and the Intel Xeon and Xeon Phi; in this work we target the Xeon and Xeon Phi architectures.

2 Kalman filter tracking

The algorithm we are targeting for parallelized processing is a Kalman filter (KF) based algorithm [4]. KF-based tracking algorithms are widely used since they are fast and incorporate estimates of multiple scattering off massive detectors directly into the trajectory of the particle. In addition, a combinatorial KF provides an elegant way to resolve ambiguities when considering multiple hits to add to a track [5]. Other track finding algorithms, more naturally suited to parallelization and coming from the image processing community, have been explored by others. These include Hough Transforms and Cellular Automata, among others (see, for instance, [6]). However, these are not the main algorithms in use at the LHC today, whereas there is extensive understanding on how KF algorithms perform. KF algorithms have proven to be robust and perform well in the difficult experimental environment of the LHC [3]. Rather than abandon the collective knowledge gained from KF algorithms, we wish to extend this robust tool by porting KF tracking to parallel architectures.

KF tracking proceeds in three main stages: seeding, building, and fitting. Seeding provides the initial estimate of the track parameters based on a few hits in a subset of layers. Realistic seeding is currently under development and will not be reported here. Building then collects additional hits in other detector layers to form a complete track, using the combinatorial KF as a basis for predicting which hits to consider and keep. Track building is by far the most time consuming step of tracking, as it requires branching to explore potentially more than one candidate track per seed after finding compatible hits on a given layer. After hits have been assigned to tracks, a final fit with a KF can be performed over each track to provide the best estimate of each tracks' parameters.

To realize performance gains, we need to exploit two types of parallelism: vectorization and parallelization. Vectorization aims to perform a single instruction on multiple data at the same time, i.e. performing the same operation across different data in lockstep. KF tracking cannot be ported in a straightforward way to run in parallel on many-core processors for the following reasons. In building, as explained above, we can branch to explore multiple candidates per seed, thus destroying the lockstep synchronization required for vectorization performance. Parallelization aims to perform different instructions at the same time on different data. The challenge to tracking then is workload balancing across different threads, as track occupancy in a detector is not uniformly distributed on a per event basis. Past work by our group has shown progress in porting sub-stages of KF tracking to support parallelism in simplified detectors (see, e.g. our presentations at ACAT2014 [7], CHEP2015 [8], and NSS-MIC2015 [9]). As the hit collection is completely determined after track building, track fitting can repeatedly apply the KF algorithm without branching, making this the ideal place to start in porting KF tracking to Xeon and Xeon Phi, with our first results shown at ACAT2014 [7].

2.1 Optimized matrix library Matriplex

The computational problem of KF-based tracking consists of a sequence of matrix operations. Unlike [4], we parameterize the measurement state and the track state in global coordinates with matrices of size $N \times N = 3 \times 3$ and $N \times N = 6 \times 6$, respectively. To allow maximum flexibility for exploring SIMD operations on small-dimensional matrices, and to decouple the specific computations from the high level algorithm, we have developed a new matrix library, MATRIPLEX. The MATRIPLEX memory layout is optimized for the loading of vector units for SIMD operations on a set of matrices. MATRIPLEX includes a code generator for defining optimized matrix operations, including support for symmetric matrices and on-the-fly matrix transposition. Patterns of elements which are known by construction to be zero or one can be specified, and the resulting generated code will be optimized accordingly to reduce unnecessary register loads and arithmetic operations. The generated code can be either standard C++ or simple intrinsic macros that can be easily mapped to architecture-specific intrinsic functions.

3 Track building

Combinatorial KF track building shares the same core KF calculations as track fitting, but has two major complications for utilizing vectorizable operations compared to track fitting. The first such problem is the "nHits" problem: when performing the KF operations, the hit set for a given candidate track in building is undefined and one potentially has to choose between $O(10^4)$ hits per layer. The second problem is the "combinatorial" problem: when more than one hit on a given layer is compatible with a given input candidate, branching occurs to explore all possible candidate tracks up to some cutoff number of candidates per seed.

The key to reducing the nHits problem is to partition the tracks and hits spatially in order to reduce the search space for a given track. We partition space without reference to the detector structure, placing hits in each layer into directionally projective bins. We first define η bins, where $\eta = -\log[\tan^{-1}(\theta/2)]$, θ being the polar angle with respect to the beam line. The η bins are self-consistent; because in our detector, tracks do not bend in η. The bins of tracks are staggered with respect to bins of hits, so that tracks never search for hits outside two overlapping hit bins when using the combinatorial KF. In this way, the number of η bins for tracks will always be an odd number. These bins provide simple boundaries for thread associations using the OpenMP parallelization library [10]. We begin with 21 η bins and distribute threads equally across the bins, such that the number of threads per η bin is either a multiple or a factor of the number of bins. We also define ϕ bins, where ϕ is the azimuthal angle, perpendicular to the beam line. As the hits are sorted in ϕ for every η bin and layer, the ϕ bins simply provide a fast look-up of compatible hits in a layer for a given track given the estimate from the KF.

With regards to the combinatorial problem, we first developed track building to add only the best hit out of all compatible hits on a layer for each track candidate. By definition, then, each seed only produces one track per layer and does not require copying of track candidates to explore multiple hits per layer, as in the full combinatorial KF approach. The best hit is defined as the hit that produced the lowest χ^2 increment to the track candidate. The vectorization and parallelization performance of this "best hit" approach were presented at CHEP2015 [8]. After demonstrating the feasibility of track building in the best hit approach, we then allowed for more than one track candidate per seed per layer to be produced, as in the full combinatorial KF approach.

3.1 Memory management

With the full combinatorial approach in place, we performed extensive studies of the performance of our software, in terms of both the physics performance and the code performance. For the latter, we used the Intel VTune 2016 [11] suite of tools to identify bottlenecks and understand the impacts of our optimization attempts. In particular, as can be expected, we determined that memory management is of critical importance. For this effect we describe below several studies to optimize memory performance, and discuss the results of these studies.

The first such impact on performance arose out significant time being spent loading data into our local caches. By using vectorizable intrinsic functions to perform copying, we saw 20% speedup. We also noticed a significant amount of time was being spent resizing the C++ vector object that stored the indices of the hits associated to a given track, and simply reserving the memory did not help. By moving to a statically sized array, the resizing was avoided, yielding a 45% speedup. The size of the data structures used in our algorithm has a crucial impact on the timing performance. In particular, the data structures for the hits and tracks must be optimized in size to fit into the fastest cache memory. We were able to reduce the size of the track objects by 20% and the hit object by 40%, leading to a 30% speedup. There were additional objects used in the algorithm that were being instantiated after each event that could instead be reset and recycled, which led to a 25% speedup. It is important to note that each of these improvements lead to individual, multiplicative speedups.

Lastly, to mitigate the impact from serial work in the copying of track candidates after finding compatible hits on each in the full combinatorial approach, we moved copying outside of all vectorizable operations in what we term the "clone engine". The clone engine approach only copies the best candidate tracks up to N candidates per seed after reading and sorting a bookkeeping list of all compatible hits. This is in contrast to a first attempt at the combinatorial KF which copied a candidate after each time a hit was deemed compatible, and sorting and keeping only the best N candidates per seed after all the possible hits on a given layer for all input candidates were explored. A more detailed discussion of this work on memory management and impacts on performance was presented at NSS-MIC2015 [9].

3.2 Results

We present here the latest vectorization and parallelization benchmarks in track building since NSS-MIC2015 given all of the developments described previously. Figure 2 contains two plots displaying the vectorization performance of three different track building approaches on Xeon, shown as functions of the number of vector units enabled. The left plot displays the absolute time in seconds for building tracks for events with 20,000 tracks as a function of the code parameter that sets the effective width of the vector unit in floats. The plot on the right shows the speedup in time compared to the time from the first data point as a function of this parameter. Only one thread is enabled in these tests.

The blue curve is the "best hit" approach described previously. Naturally, this approach will have the lowest absolute time in comparison to the two fully combinatorial approaches, the black and pink curves. The black curve is the original approach to combinatorial track finding where the copying of the track candidates is inside the vectorizable KF operations. The pink curve shows the clone engine approach, which moves the serial work of copying outside of the KF vectorizable operations. As expected, the clone engine approach has a lower absolute time and higher speedup than the original. It is important to note that while the best hit approach is the fastest, the physics performance has inefficiencies in hit finding and track finding from not being fully able to explore multiple track candidates per seed. Even in our simplified model, this behavior is already apparent, and the best hit approach is expected to become even more inefficient as we add in more realistic detector effects.

Figure 3 shows the same vectorization plots as Figure 2, now with Xeon Phi, which has AVX-512 vector registers that are twice as wide as those in Xeon. The original combinatorial approach is not shown, as it was apparent from Figure 2 that it did not perform as well as the clone engine approach. With one thread enabled, Xeon Phi sees the same 1.5× to 2× speedup as Xeon. There is a noticeable

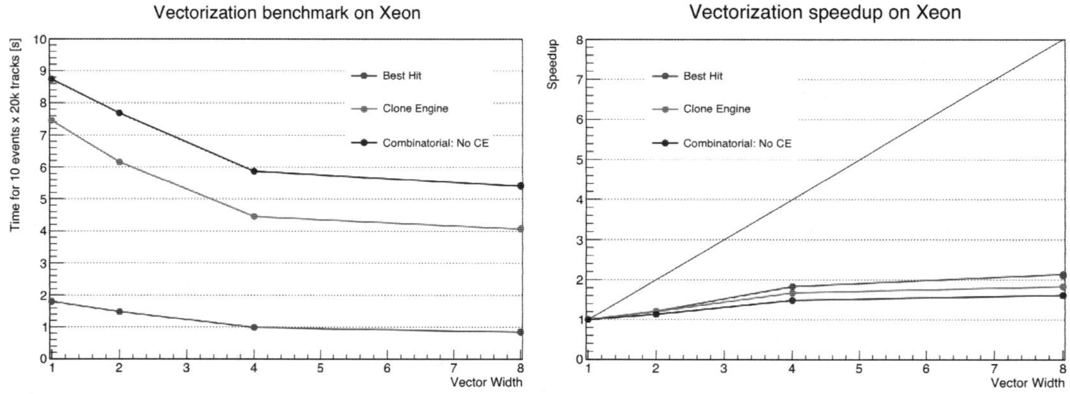

Figure 2. Left: Absolute time as a function of the code parameter that sets the effective vector width. Right: Speedup as a function of this parameter on Xeon. Both plots show results for three different track building approaches. The blue curve is the no combinatorial tracking approach. The black and pink curves use the full combinatorial approach in track building. The pink curve moves the copying of tracks outside of the vectorizable operations. Only one thread is enabled.

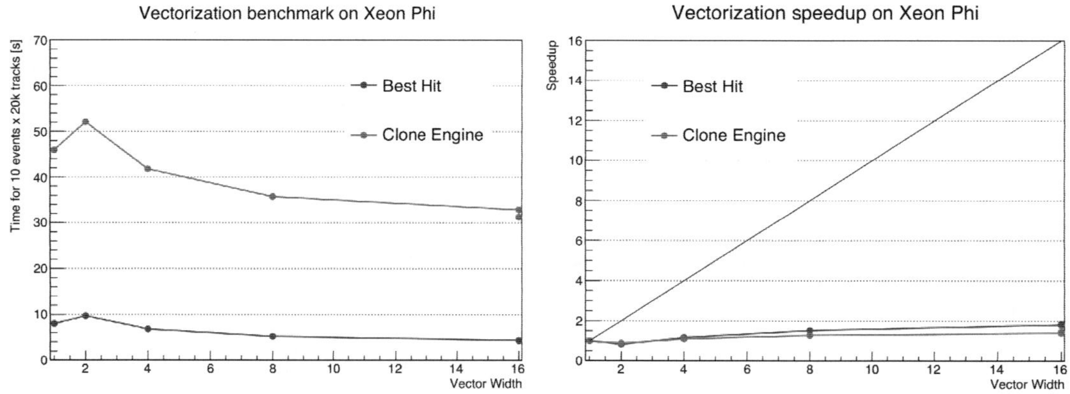

Figure 3. Left: Absolute time as a function of the code parameter that sets the effective vector width on Xeon Phi. Right: Speedup as a function of this parameter on Xeon Phi. Both plots show results for two different track building approaches. The blue curve is the no combinatorial tracking approach. The pink curve uses the full combinatorial approach in track building and moves the copying of tracks outside of the vectorizable operations. Only one thread is enabled.

increase in time after the vector unit size is first doubled. This is likely overhead from enabling the vector units. As seen on Xeon, the best hit approach still has the lowest absolute timing.

Figure 4 displays the parallelization performance in terms of the absolute time and speedup of the clone engine approach on Xeon as a function of the number of threads enabled using OpenMP v. 4.0. The best hit approach is removed from these plots, because in reality, the full combinatorial approach is one we will ultimately use. The Xeon machine we are using has 12 physical cores which appear to be 24 logical cores to the OS due to hyperthreading being enabled. As described previously in Section 3, we enable threads as multiples or factors of the number of η bins. Since we start with 21 η

 DOI: 10.1051/epjconf/201612700010

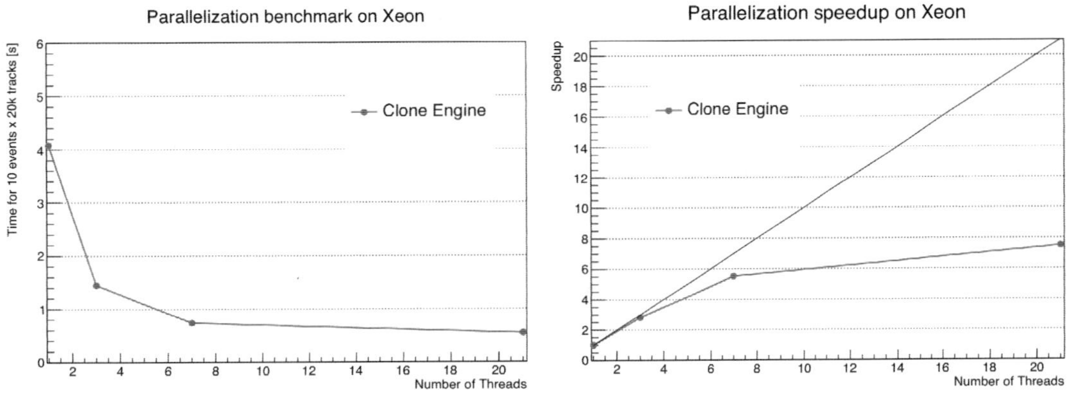

Figure 4. Left: Absolute time as a function of the number of threads enabled on Xeon. Right: Speedup as a function of the number of threads enabled on Xeon. Both plots show results for the full combinatorial approach in track building that moves the copying of tracks outside of the vectorizable operations. The full Xeon vector width (8 floats) is assumed by the code.

bins, we cannot explore the critical boundary of 12 threads with OpenMP. The slight deviation away from ideal scaling before hyperthreading arises from the fact that some threads are unlucky and end up with more work per event than the rest, stalling the other threads, which was shown with VTune. The large deviation away from ideal scaling after enabling hyperthreading is due to resource limitations: the two threads per core are contending for the same instruction pipelines and data caches. Even so, nearly 8× speedup is seen for 21 threads.

Figure 5 displays the same parallelization performance plots as Figure 4, now in terms of the Xeon Phi, with the full vector width of 16 floats enabled using OpenMP v. 4.0. The Xeon Phi we are using has 61 physical cores requiring 122 independent instruction streams for full utilization, due to the fact that the Xeon Phi issues instructions for a given thread every other clock cycle. Therefore, to keep a given physical core busy every clock cycle, one has to schedule two threads per core alternating in instruction execution. A form of hyperthreading is also present on Xeon Phi, yielding a total of 244 hardware threads (logical cores). Thus we notice two features in the speedup curve for Xeon Phi, occurring just past the physical core count (61) and twice that number (122). However, this loss in speedup is eventually recovered, and a factor of 30× speedup is observed with 210 threads enabled.

4 Conclusion and outlook

We have made significant progress in parallelized and vectorized Kalman filter-based end-to-end tracking R&D on Xeon and Xeon Phi architectures. Through the use of a wide toolkit which includes Intel VTune, we have developed a good understanding of bottlenecks and limitations of our implementation which has led to further improvements. We are beginning to explore Intel Threaded Building Blocks (TBB), which allows for more dynamic scheduling to prevent stalls in resource usage. In this way, threads can be spawned in numbers more naturally suited to the number of cores, unlike OpenMP. The project has seen promising initial results; however, much work remains.

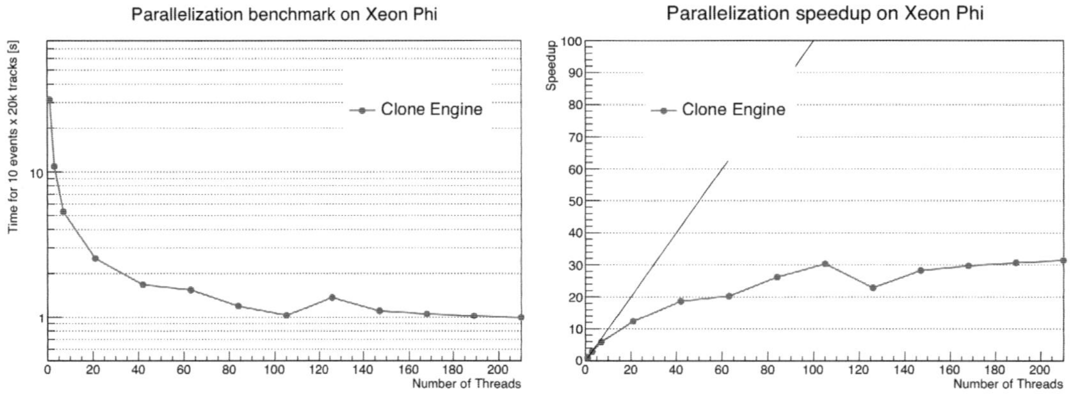

Figure 5. Left: Absolute time as a function of the number of threads enabled on Xeon Phi. Right: Speedup as a function of the number of threads enabled on Xeon Phi. Both plots show results for the full combinatorial approach in track building that moves the copying of tracks outside of the vectorizable operations. The full Xeon Phi vector width (16 floats) is assumed by the code.

Acknowledgment

This work is supported by the U.S. National Science Foundation, under the grants PHY-1520969, PHY-1521042, PHY-1520942 and PHY-1120138.

References

[1] CMS Collaboration, Phys. Lett. **B716**, 30 (2012)

[2] ATLAS Collaboration, Phys. Lett. **B716**, 1 (2012)

[3] G. Cerati, Proceedings of Science **VERTEX2014**, 037 (2015), `http://pos.sissa.it/archive/conferences/227/037/Vertex2014_037.pdf`

[4] R. Frühwirth, Nucl. Inst. & Meth. A **262**, 444 (1987)

[5] P. Billoir, Comp. Phys. Comm. **225**, 352 (1984)

[6] V. Haylo, P. LeGresley, P. Lujan, V. Karpusenko, and A. Vladimirov, JINST **9**, P04005 (2014)

[7] G. Cerati et al., J. Phys. Conf. Ser. **608**, 012057 (2014)

[8] G. Cerati et al., J. Phys. Conf. Ser. **664**, 072008 (2015)

[9] G. Cerati et al., *Kalman-filter-based particle tracking on parallel architectures at hadron colliders* (2015), `arXiv:1601.08245`

[10] OpenMP Architecture Review Board, *Openmp application program interface version 4.0* (2013), `http://www.openmp.org/mp-documents/OpenMP4.0.0.pdf`

[11] See `https://software.intel.com/en-us/intel-vtune-amplifier-xe`, accessed 23 March 2016

Graphics Processors in HEP Low-Level Trigger Systems

Roberto **Ammendola**[1], Andrea **Biagioni**[2], Stefano **Chiozzi**[5], Angelo **Cotta Ramusino**[5], Paolo **Cretaro**[2], Stefano **Di Lorenzo**[3,4], Riccardo **Fantechi**[3], Massimiliano **Fiorini**[5,6], Ottorino **Frezza**[2], Gianluca **Lamanna**[8], Francesca **Lo Cicero**[2], Alessandro **Lonardo**[2], Michele **Martinelli**[2], Ilaria **Neri**[5], Pier Stanislao **Paolucci**[2], Elena **Pastorelli**[2], Roberto **Piandani**[3], Luca **Pontisso**[3,a], Davide **Rossetti**[7], Francesco **Simula**[2], Marco **Sozzi**[3,4], and Piero **Vicini**[2]

[1] *INFN Sezione di Roma Tor Vergata, Italy*
[2] *INFN Sezione di Roma, Italy*
[3] *INFN Sezione di Pisa, Italy*
[4] *Università di Pisa, Italy*
[5] *INFN Sezione di Ferrara, Italy*
[6] *Università di Ferrara, Italy*
[7] *NVIDIA Corp., Santa Clara, CA, USA*
[8] *INFN Laboratori Nazionali di Frascati, Italy*

Abstract. Usage of Graphics Processing Units (GPUs) in the so called general-purpose computing is emerging as an effective approach in several fields of science, although so far applications have been employing GPUs typically for offline computations. Taking into account the steady performance increase of GPU architectures in terms of computing power and I/O capacity, the real-time applications of these devices can thrive in high-energy physics data acquisition and trigger systems. We will examine the use of online parallel computing on GPUs for the synchronous low-level trigger, focusing on tests performed on the trigger system of the CERN NA62 experiment. To successfully integrate GPUs in such an online environment, latencies of all components need analysing, networking being the most critical. To keep it under control, we envisioned NaNet, an FPGA-based PCIe Network Interface Card (NIC) enabling GPUDirect connection. Furthermore, it is assessed how specific trigger algorithms can be parallelized and thus benefit from a GPU implementation, in terms of increased execution speed. Such improvements are particularly relevant for the foreseen Large Hadron Collider (LHC) luminosity upgrade where highly selective algorithms will be essential to maintain sustainable trigger rates with very high pileup.

1 Introduction

In High Energy Physics experiments the realtime selection of the most interesting events is of paramount importance because of the collision rates which do not give the possibility to save all the data for offline analysis. For this purpose, different trigger levels are usually used to carefully choose the most meaningful events. Low levels require low and (almost) deterministic latency, and their

[a] e-mail: luca.pontisso@pi.infn.it

standard implementation is on dedicated hardware (ASICs or FPGAs). Our approach aims at exploiting the Graphics Processing Units (GPUs) computing power, in order to build refined physics-related trigger primitives, such as energy or direction of the final state particles in the detectors, and therefore leading to a net improvement of trigger conditions and data handling. GPUs architectures are massively parallel, being designed to optimize computing throughput but with no particular attention to their usage in real-time contexts, such as the online low-level triggers. While execution times are rather stable on these devices, also I/O tasks have to guarantee real-time features along the data stream path, from detectors to GPU memories. The NaNet project arises with the goal of designing a low-latency and high-throughput data transport mechanism for systems based on CPU/GPUs. The GPU-based low-level trigger using the NaNet board is currently integrated in the experimental setup of the Ring-imaging Čerenkov detector (RICH) of the NA62 experiment in order to reconstruct the ring-shaped hit patterns. We report and discuss results obtained with this system along with the algorithms that will be implemented.

2 NaNet

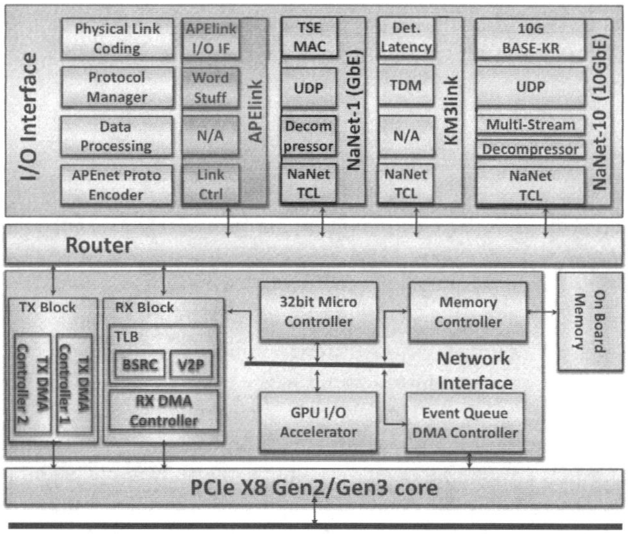

Figure 1. NaNet PCIe Network Interface Card family architecture.

The NaNet project goal is the design and implementation of a family of FPGA-based PCIe Network Interface Cards, equipped with different network links according to their employment, to bridge the detectors' front-end electronics and the computing nodes [1]. To this pourpose NaNet features (i) a low-latency and high-throughput data transport mechanism for real-time systems, (ii) support for several link technologies to be able to fit in different experimental setups, and (iii) a design with multi-port interface to increase the scalability of the entire system to reduce the number of nodes in the farm clusters. The management of custom and standard network protocols is performed in hardware to avoid jitter effects (variance in tasks' execution time) due to operating system and to guarantee a deterministic behaviour of communication latency while achieving the maximum capability out of the adopted channel. Furthermore, NaNet integrates a processing stage which is able to process data coming from detectors on the fly, to improve the efficiency of applications running on computing nodes.

 DOI: 10.1051/epjconf/201612700011

On a per-experiment basis, different solutions can be implemented: data decompression, data reformatting, merging of event fragments. Finally, data transfers to or from application memory in CPU or GPU are performed with zero-copy. The block diagram of the NaNet architecture is in Figure 1. Exploiting the modularity of the hardware design, the Network Interface and Router modules are shared among all boards of the family, ensuring fast and reliable implementation of different NaNet design configurations. The PCIe interface can be customized — *i.e.* Gen2 or Gen3 — and the number or switch ports chosen, to best match experimental requirements. The I/O interface, one of the key features of the NaNet design, consists of 4 elements: Physical Link Coding, Protocol Manager, Data Processing, APEnet Protocol Encoder. The Physical Link Coding covers the Physical and DataLink Layers of the Open Systems Interconnection (OSI) model, managing transmission of data frames over a common local media and performing error detection. The Protocol Manager covers the Network and Transport Layers, managing the reliability of communication data flow control. A processing stage, Data Processing, that applies some function to the data stream in order to offload some work for the applications running on the computing node. Finally, the APEnet Protocol Encoder performs a protocol translation to a format more suited for PCIe Direct Memory Access (DMA) transactions.

NaNet accomplishes zero-copy networking by means of a hardware implemented RDMA (Remote DMA) engine working with both CPUs and nVIDIA GPUs (GPUDirect V2/RDMA [2]), achieving very low I/O latency.

On the host side, a dedicated Linux kernel driver offers its services to an application-level library, which provides the user with a series of functions to: open/close the NaNet device; register and de-register circular lists of persistent data receiving buffers (CLOPs) in GPU and/or host memory; manage software events generated when a receiving CLOP buffer is full (or when a configurable timeout is reached) and received data are ready to be processed.

NaNet-1 was developed in order to verify the feasibility of the project; it is a PCIe Gen2 x8 network interface card featuring GPUDirect RDMA over GbE.

NaNet-10 is a PCIe Gen2 x8 network adapter implemented on the Terasic DE5-net board equipped with an Altera Stratix V FPGA featuring four enhanced small form-factor pluggable (SFP+) cages [3].

Both implementations use UDP as their transport protocol.

In Figure 2, NaNet-10 and NaNet-1 latencies are compared within UDP datagram size range; NaNet-10 guarantees sub-μs hardware latency for buffers up to ≈1 kB in GPU/CPU and it reaches its 10 Gbit/s bandwidth peak capability already at ≈1 kB size (Figure 3).

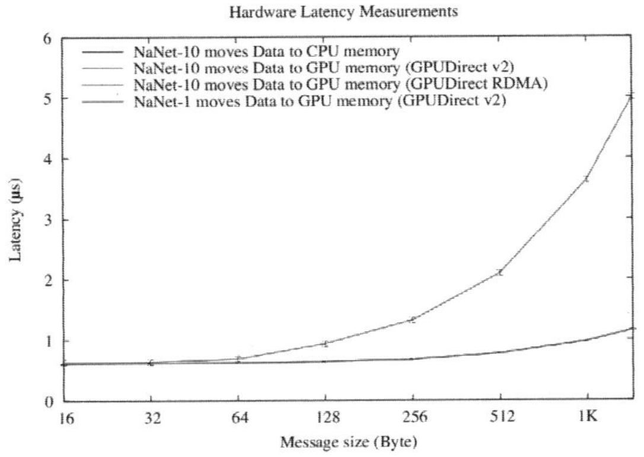

Figure 2. NaNet-10 vs. NaNet-1 hardware latency. NaNet-10 curves are completely overlapped at this scale.

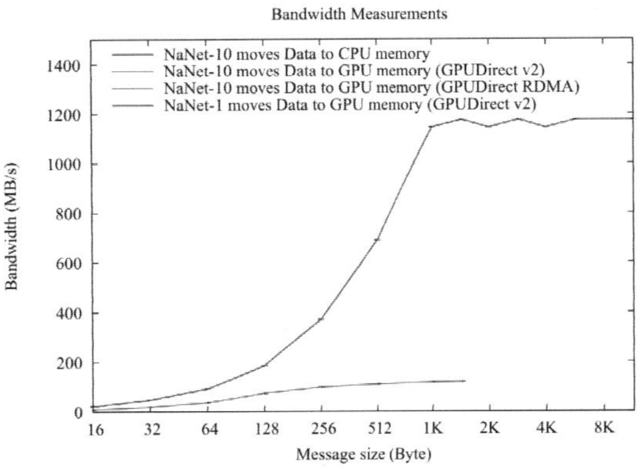

Figure 3. NaNet-10 vs. NaNet-1 bandwidth. NaNet-10 curves are completely overlapped at this scale.

DOI: 10.1051/epjconf/201612700011

3 GPU-based L0 trigger for the NA62 RICH detector

In order to research the feasibility of a GPU-based L0 trigger system for NA62 experiment, we directed our attention on reconstructing the ring-shaped hit patterns in the RICH of the NA62 experiment. This detector identifies pions and muons with momentum in the range between 15 GeV/c and 35 GeV/c. Čerenkov light is reflected by a composite mirror with a focal length of 17 m focused onto two separated spots equipped with $\approx$ 1000 photomultipliers (PM) each. Data communication between the readout boards (TEL62) and the L0 trigger processor happens over multiple GbE links using UDP streams. The final system consists of 4 GbE links to move primitives data from the readout boards to the GPU_L0TP (see Figure 4). The overall time budget for the low-level trigger comprising both communication and computation tasks is of 1 ms, so a deterministic response latency from GPU_L0TP is a strict requirement. Refined primitives coming from the GPU-based calculation will be then sent to the central L0 processor, where the trigger decision is made taking in account informations from other detectors.

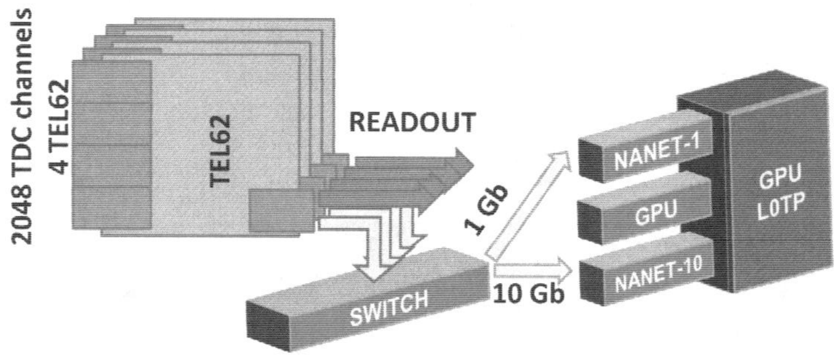

Figure 4. Pictorial view of GPU-based trigger.

3.1 Reconstructing multiple ring events on GPUs

The knowledge of Čerenkov rings parameters allows to build stringent conditions for data selection at trigger level. This implies that circles have to be reconstructed using the coordinates of activated PMs. We take in consideration two multi-rings pattern recognition algorithms based only on geometrical considerations (no other information is available at this level) and particularly suitable for exploiting the intrinsic parallel architecture of GPUs.

3.1.1 Histogram algorithm

The procedure involves dividing the xy-plane into a grid and creating a histogram whose bins contain distances from the gridpoints and the hits of the physics event. Distance bins whose contents exceed a threshold value let identify the rings. In order to limit the use of resources, it is possible to proceed in two steps, starting the histogram procedure with a 8x8 grid and calculating now distances from such squares. Afterwards, to refine their positions, the calculation is repeated with a grid 2x2 only for the candidate centers selected according to the threshold in the previous step.

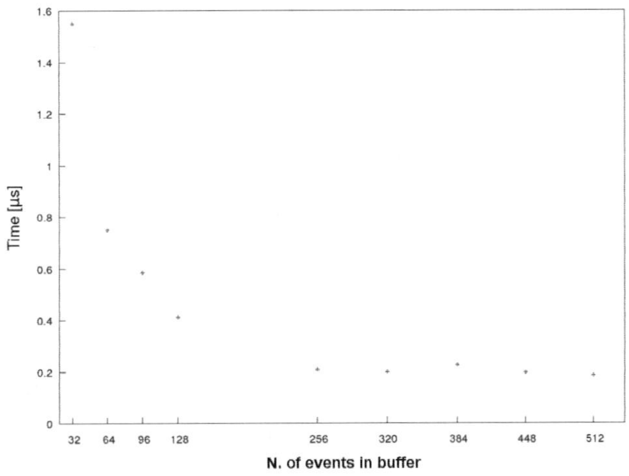

Figure 5. Almagest algorithm performances, time for single event. Test performed on K20c nVIDIA GPU.

3.1.2 Almagest algorithm

One of the Ptolemy's theorems described in the *Almagest*, his treatise on astronomy, states that when four vertices of a quadrilateral (ABCD) lie on a common circle, it is possible to relate four sides and two diagonals: $|AC| \times |BD| = |AB| \times |CD| + |BC| \times |AD|$. By using this formula it is possible to implement a pattern recognition algorithm for multi-rings which exposes different level of parallelism, resulting well-suited for GPUs achitecture and fast in its execution. This is crucial either to directly reconstruct the rings or to choose different algorithms according to the number of circles. The large number of possible combinations of four vertices, given a maximum of 64 points for physics event, can be a limitation to this approach. To greatly reduce the number of tests, one possibility is to choose few triplets — *i.e.* a set of three hits assumed to belong to the same ring — trying to maximize the probability that all their points belong to the same ring and iterating through all the remaining hits to search for the ones satisfying the aforementioned formula [4]. The parallel implementation of this algorithm yields many triplets and events being processed at the same time. Some results are shown in Figure 5, where the computing time is further sped up by greatly reducing accesses to GPU shared memory, mainly using threads private registers through CUDA (a parallel computing platform and programming model invented by nVIDIA) intra-warp veto and shuffle instructions, so that multi-rings events are processed with a latency of 0.5 µs per event. Once the number of rings and points belonging to them have been found, it is possible to apply for example Crawford's method [5] to obtain centre coordinates and radii with better spatial resolution.

4 Results for the GPUbased L0 trigger with NaNet-1

In 2015 the GPU-based trigger at CERN includes 2 TEL62 boards connected to a HP2920 switch and a NaNet-1 [6] board with a TTC HSMC daughtercard plugged into a server made of a X9DRG-QF dual socket motherboard populated with Intel Xeon E5-2620 @2.00 GHz CPUs (*i.e.* Ivy Bridge architecture), 32 GB of DDR3 RAM and a Kepler-class nVIDIA K20c GPU.

Such a system allows testing of the whole chain: the data events move towards the GPU-based trigger through NaNet-1 by means of the GPUDirect RDMA interface. All data landing within a

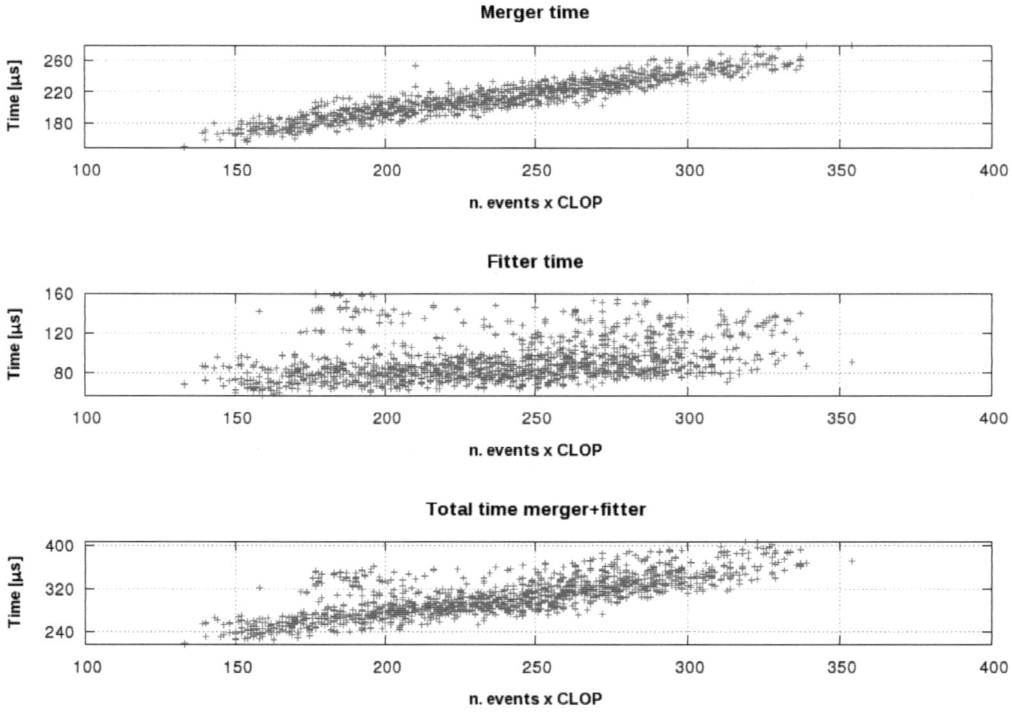

Figure 6. Multi-ring reconstruction of events performed on K20c nVIDIA GPU.

configurable time frame are gathered and organized in a parametrizable Circular List Of Persistent buffers (CLOP) in GPU memory. This time frame must be always shorter or equal to how long multi-ring reconstruction takes on the GPU, to be sure that buffers are not overwritten before they are consumed.

Results are reported in Figure 6. On the x-axis: the CLOP size measured as number of received events; on the y-axis: the latencies of different stages. The computing kernel implemented the histogram fitter with a single step (*i.e.* using an 8x8 grid only). Events coming from two readout boards, for a gathering time of 400 μs, and parameters like events rate (collected with a beam intensity of 4×10^{11} protons per spill), a CLOP's size of 8 kB, time frame were chosen so that we could test the online behaviour of the trigger chain. The merge operation doesn't expose much parallelism, requiring instead synchronization and serialization. As a result it is an ill-suited problem to the GPU architecture. In operative conditions, the merging time only would exceed the time frame. The high latency of the this task when performed on a GPU suggests to offload such duty to a hardware implementation.

Acknowledgements

S. Chiozzi, A. Cotta Ramusino, S. Di Lorenzo, R. Fantechi, M. Fiorini, I. Neri, R. Piandani, L. Pontisso, M. Sozzi thank the GAP project, partially supported by MIUR under grant RBFR12JF2Z "Futuro in ricerca 2012".

References

[1] A. Lonardo et al., JINST **10**, C04011 (2015)

[2] R. Ammendola et al., *GPU Peer-to-Peer Techniques Applied to a Cluster Interconnect*, in *Proceedings of the IEEE 27th International Parallel and Distributed Processing Symposium Workshops and PhD Forum (IPDPSW 2013)* (2013), pp. 806–815, http://ieeexplore.ieee.org/stamp/stamp.jsp?tp=&arnumber=6650959

[3] R. Ammendola et al., JINST **11**, C03030 (2106)

[4] G. Lamanna, Nucl. Inst. & Meth. A **766**, 241 (2014)

[5] J. Crawford, Nucl. Inst. & Meth. **211**, 223 (1983)

[6] R. Ammendola et al., JINST **9**, C02023 (2104)

EPJ Web of Conferences **127**, 00012 (2016) DOI: 10.1051/epjconf/201612700012
Connecting The Dots 2016

A Proposal for the Upgrade of the Muon Drift Tubes Trigger for the CMS Experiment at the HL-LHC

Nicola **Pozzobon**[1,2,a], Pierluigi **Zotto**[1,2], and Fabio **Montecassiano**[2]

[1] *Università degli Studi di Padova, Italy*
[2] *INFN Sezione di Padova, Italy*

Abstract. A major upgrade of the readout and trigger electronics of the CMS Drift Tubes muon detector is foreseen in order to allow its efficient operation at the High Luminosity LHC. A proposal for a new L1 Trigger Primitives Generator for this detector is presented, featuring an algorithm operating on the time of charge collection measurements provided by the asynchronous readout of the new TDC system being developed. The algorithm is being designed around the implementation in state-of-the-art FPGA devices of the original development of a Compact Hough Transform (CHT) algorithm combined with a Majority Mean-Timer, to identify both the parent bunch crossing and the muon track parameters. The current state of the design is presented along with the performance requirements, focusing on the future developments.

1 Introduction

Muon identification at the CMS experiment relies on a redundant detection system exploiting three different technologies: Drift Tube chambers (DT) in the central region $|\eta| < 1.2$, Cathode Strip Chambers (CSC) in the forward region $|\eta| > 1.1$ and Resistive Plates Chamber (RPC) in both regions [1]. Signals from all these sub-systems are used to build the Level 1 Trigger (L1) decision for the on-line identification of muon tracks. However, the currently available on-line momentum resolution does not allow to retain manageable trigger rates after the High Luminosity LHC upgrade is completed [2]. Several upgrades of the muon electronics are under design, including the replacement of the current DT readout and L1 local trigger. The current L1 DT local trigger electronics was produced in the early 2000 and is hosted on-detector, therefore it is old and suffering from radiation damage. Hence a project aiming to the full replacement of the readout and trigger electronics was started. Given the limited amount of information to be moved, the current design of the new readout foresees an asynchronous data feed via high speed optical link of all collected signals, straight from the detector, without any intermediate decision stage. The DT local Trigger Primitive Generator (TPG) will then need to operate on time-stamped data stored in large memories off-detector. This approach is not compatible with the current DT TPG [3], which is based on a synchronous track segment reconstruction designed around a time-dependent fitter for the parent bunch crossing identification. Therefore, a new, modern, fast and efficient trigger scheme needs to be developed.

[a]e-mail: nicola.pozzobon@pd.infn.it

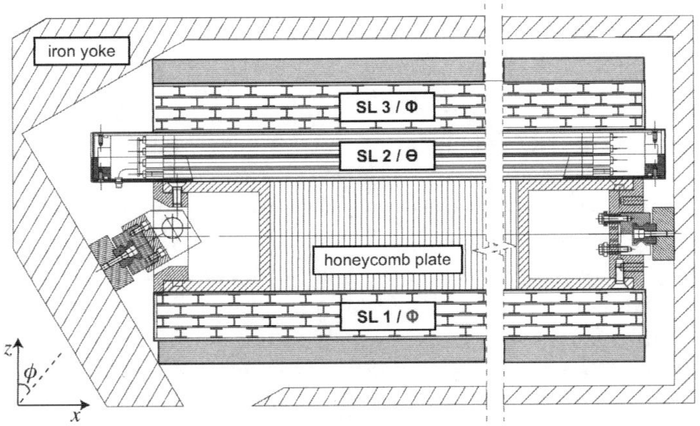

Figure 1. Layout of a CMS DT chamber. Wires are arranged in layers, and parallel layers are grouped four-by-four into super-layers. Each chamber features two super-layers in the $r\phi$ projection and one super-layer in the θ projection. Each drift cell is 4.2 cm wide and 1.3 cm high, while the distance d between the centre of the two $r\phi$ super-layers is 23.5 cm. The θ super-layer is not used in this preliminary study.

2 Requirements

The 240 CMS DT chambers, interleaved with the CMS solenoid return iron yoke, can be described as telescopes of position-sensitive detectors, grouped together in parallel layers, as shown in Figure 1. Each chamber provides an array of charge collection times t_i measured from a common starting signal, called BC0, distributed by the machine and corresponding to the first bunch crossing in an LHC cycle. These times are converted into drift time measurements T_i, once the parent bunch crossing time t_p with respect to BC0 is identified.

The new trigger scheme must use the drift-times in order to identify approximately straight track segments[1] with an efficiency and a resolution comparable to the ones of the current trigger system: the required efficiency must be higher than 98% over a range spanning approximately $\pm 50°$ in the local angle ϕ, the angular resolution must be better than 30 mrad (3 mrad) using measurements from 4 (8) wires, and the resolution on track intercept must be better than 1.4 mm.[2]

Each drift time T_i is converted into a drift distance to the anode wire, but each reconstructed muon hit in any drift time chamber comes accompanied by a second position measurement due to the unresolved left-right ambiguity. Therefore, the track segment reconstruction must deal with an amount of spurious position measurements of the same order of the actual crossing coordinates of the muon in addition to the noise contribution. The trigger must then provide robust identification of the muon within a non negligible amount of background with only $O(10)$ measurements and limiting the amount of ghost tracks in order to keep rates at a manageable level, within few μs latency.

The proposed approach is split in two parts: the determination of the parent bunch crossing time t_p by means of a mean-timing technique and the evaluation of the muon track parameters by means of a fast and dedicated implementation of the Hough transform (HT). We will first discuss the HT

[1]The magnetic field is confined in the iron yoke: the stray field in the chambers has negligible effects in most positions [4].

[2]The track intercept is defined as the intersection of the reconstructed track segment with a reference plane. The reference plane corresponding to the quoted reference value is the one parallel to $r\phi$ super-layers at equal distance from both. The problem of an optimal definition of such reference plane is outlined in Section 3 and described in detail in [5].

implementation assuming t_p is exactly known, and afterwards introduce the t_p computation method in order to decouple algorithmic effects.

3 Hough transform for the DT TPG

The Hough transform [6–8] is a well-established pattern recognition technique worth being investigated for a new DT TPG, given its robustness and simplicity, but whose implementation requires a lot of computing resources. We performed a first evaluation study to identify and tailor a HT implementation scheme that can match the efficiency and resolution requirements listed in Section 2 and which could be suitable for implementation in programmable logic devices. We have shown that a parameter reduction, combined with a clever partitioning of the input dataset and dimensioning of the HT histograms, brings the size of HT histograms and the complexity of the computations compatible with the typical specifications of state-of-the-art FPGA devices [5][3] . In this contribution we report only on the status of our studies for the identification of track segments in the $r\phi$ projection, therefore we refer to measurements of super-layers 1 (inner) and 3 (outer) according to the convention of Figure 1.

The layout of the CMS DT chambers intrinsically accounts for non-degenerate slope-intercept description using the reference system shown in Figure 1. A straight track is therefore described by the equation:

$$x = -m \cdot z + c,$$

where the local track angle $\phi = \tan^{-1} m$ is measured with respect to the orthogonal direction $\hat{z}$. For each recorded time, the position of the layer defines the z coordinate and x is the transverse coordinate reconstructed from the drift time. In the HT parameter space, each (x, z) measurement is transformed into a line of slope z and intercept x:

$$c = z \cdot m + x. \tag{1}$$

These lines fill the bins of a two-dimensional histogram in the (m, c) parameter space and intersect when the same parameters are found for different (x, z) points, thus originating an accumulation in the parameter space. The best parameters values are identified as the position within the parameter space where the majority of contributions is gathering.

The actual algorithm implementation features three independent HT histograms which are filled in parallel: one for super-layer 1, one for super-layer 3, and one combining both of them together. The histograms are then compared to each other in order to define the best values of muon track parameters. The DT drift cells in each super-layer are grouped into macro-cells spanning approximately 16 cm in x, to reduce the number of input hits to be considered simultaneously for track identification, and partially overlapped, to ensure optimal redundancy and hermeticity for track finding up to $|\phi| \approx 50°$.

The coordinate z of each measured point becomes the slope of the corresponding line in the parameter space. Such transformed lines can become degenerate with large values of z, i.e. almost vertical and separated by very small angles. This can affect the identification of the peak in the parameter space, which is better identified by lines crossing each other with large and evenly distributed angles. This condition corresponds to having measurements evenly distributed around $z = 0$. Hence, when using only one super-layer, we set $z = 0$ between its central layers, while in case both super-layers are used, we set $z = 0$ on the median plane of the chamber.

Therefore, for a given track, the input measurements are naturally split into three different subsets, allowing to identify the track by means of three different HT histograms (an *inner HT* for super-layer 1; an *outer HT* for super-layer 3; a *chamber HT* for both super-layers) that are recombined with each other after histogram filling completion. Histograms are recombined on a track-slope basis, being this

[3]Patent pending on key features of the algorithm.

 DOI: 10.1051/epjconf/201612700012

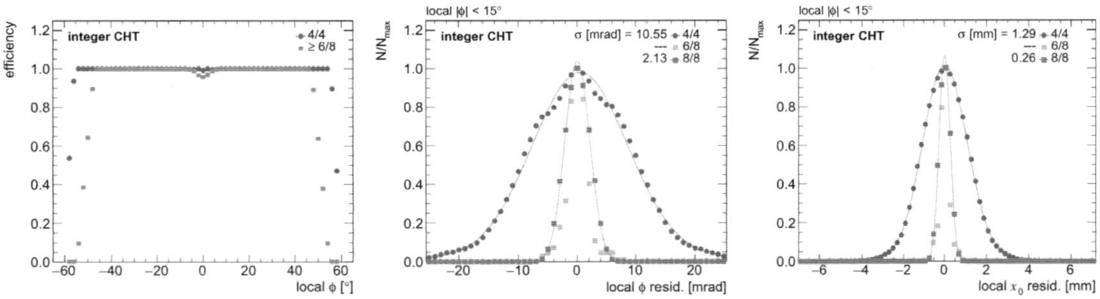

Figure 2. CHT algorithm efficiency for finding track segments with 4 aligned reconstructed hits in one super-layer (4/4) and for at least 6 aligned reconstructed hits in both super-layers ($\geqslant 6/8$), shown as a function of the local track angle ϕ (left). Resolution of the reconstructed local track angle ϕ (centre) and intercept x_0 (right) with 4 aligned reconstructed hits in one super-layer (4/4), with 6 aligned reconstructed hits in both super-layers (6/8), and with 8 aligned reconstructed hits in both super-layers (8/8), for tracks with local angle $|\phi| < 15°$. The resolution is expressed as the square root of the variance of a weighted sum of two normal functions used to fit the distribution of the residuals. The muon crossing time t_{p} is assumed to be exactly known at this stage, to decouple the bunch crossing identification effects.

parameter non sensitive to the choice of $z = 0$. In a Standard HT (SHT) approach, this means that histograms are projected onto the m- axis after a pre-defined threshold on the number of entries is applied to the parameter space bins. Different thresholds are used to assign a quality to the candidate track segment, the quality being higher for higher counts. Such projected histograms produce m-bitsets, which are then combined together with a bitwise 'AND' operation in order to identify the smallest subset of slope values common to all the three subsets. The best m value is then obtained by means of a simple clustering algorithm.

The track intercept $c_{\mathrm{inn,out}}$ is recalculated for each single super-layer subset and afterwards converted to the value x_0 in the chamber reference frame. If both super-layers are available, the estimate of m can be improved exploiting the large lever arm:

$$x_0 = \frac{c_{\mathrm{inn}} + c_{\mathrm{out}}}{2},$$
$$m = \frac{c_{\mathrm{inn}} - c_{\mathrm{out}}}{d}.$$

If, on the contrary, only one super-layer returns a positive result, m is unchanged from the low resolution m-bitset estimate, and x_0 is the extrapolation of the local intercept to the middle plane of the chamber.

A Compact HT (CHT) approach was developed in order to reduce the number of parameters. The parameter reduction exploits the fact that each pair of aligned hits must return the same values for m and c. Given that only m is used in the histogram recombination scheme, one can consider pairs of measurements and combine Equation (1) for two different measurements into

$$\Delta c = \Delta z \cdot m + \Delta x,$$

being Δx and Δz the difference in local coordinates between the considered reconstructed pair of positions. In a HT mapping Δc versus m, Δc is expected to accumulate at zero for all pairs of hits actually belonging to the same track, given that the chosen bin size is large enough to account for the effects of uncertainties in position measurements. Therefore, only the histogram row corresponding to $\Delta c = 0$ is relevant for the determination of hit alignment. This allows to handle one-dimensional histograms, i.e. only the $\Delta c \approx 0$ row of a two-dimensional $(m, \Delta c)$ histogram. Defining the size of the bins and the best thresholds requires an accurate tuning of the available options. The best working points identified in our study indicate a size reduction of the HT histogram area, comparing the SHT and the CHT approach, by a factor ≈ 35. The benefit on the histogram size of moving from the use of single hits of a SHT to couples of hits has the drawback of a more complicated handling of input hits.

Efficiency and resolution for different qualities were evaluated with a fully-integer representation of the algorithm and are shown in Figure 2, assuming the muon crossing time is exactly known. A first Register Transfer Language (RTL) high-level synthesis of the CHT algorithm has been performed showing that it can be implemented in a state-of-the-art FPGA, although the timing performance is still far from requirements. Variants of the method improving the latency are being studied.

4 Bunch crossing identification

The current bunch crossing (BX) identification scheme is based on a time-dependent synchronous fitter (the BTI ASIC [9]), sampling different BX hypotheses with 12.5 ns precision. Since this scheme cannot work with the foreseen asynchronous data feed, we propose another way to exploit the mean-timing, which is one of the key features of the CMS DT chambers design since RD5 [10]. Any meaningful triplet of wires (labeled 'A', 'B', and 'C'), in different layers within a super-layer and assuming a left-right hypothesis in each tube, is associated to a linear combination $\mathcal{L}$ of the drift times T_A, T_B, T_C and the maximum drift time in a tube T_{max}:

$$\mathcal{L}(T_{max}, T_A, T_B, T_C) = 0, \tag{2}$$

where all coefficients in $\mathcal{L}$ are integer.[4] One such example is presented in Figure 3. The very same linear combination can be made explicit in terms of the particle crossing time t_p, inverting the relation between the drift time T_i and the time to charge collection times t_i

$$\mathcal{L}(T_{max}, t_A, t_B, t_C) - S_{ABC} \cdot t_p = 0, \tag{3}$$

being the coefficients of the linear combination $\mathcal{L}$ identical in (2) and (3) for corresponding patterns, and S_{ABC} the algebraic sum of the coefficients multiplying T_A, T_B, T_C in (2). The majority of the meaningful patterns has $S_{ABC} \neq 0$, making it straightforward to compute t_p at once for those patterns.

All meaningful patterns in a macro-cell are used to compute t_p by means of tables and counts, without the need of t_p sampling as done in the BTI. The most voted t_p is chosen and used to convert each measured t_i into a pair of reconstructed positions $x_{i,left}$ and $x_{i,right}$. If the t_p computed from the inner super-layer macro-cell is different from the one computed in the outer super-layer macro-cell, both hypotheses are tested: typically, only the good one returns a valid track, since the CHT histogram built from both super-layers naturally suppresses ghost track segments. The procedure can be then defined as a static Majority Mean-Timer (MMT). The performance of the MMT+CHT chain has been evaluated with a fully-integer representation of both algorithms, assigning each track a random BX over one LHC orbit, and the relevant figures of merit are shown in Figure 4, without any check on the MMT result, and in Figure 5, where only in-time muon candidates are plotted. A comparison between the two Figures shows that the evaluation technique does not introduce any resolution degradation, while introducing some out-of-time and therefore wrong triggers. Figure 5 shows also that the mean-timer is not working beyond $|\phi| \approx 50^\circ$ where the contribution of single super-layers is crucial, while all in-time evaluations provide excellent resolution. The fraction of wrong t_p identifications at angles $|\phi| < 40^\circ$ is of the order of 10^{-4}.

[4]This is due to the specific layout of the CMS DT chambers, where tubes have all the same height and are staggered by half tube width.

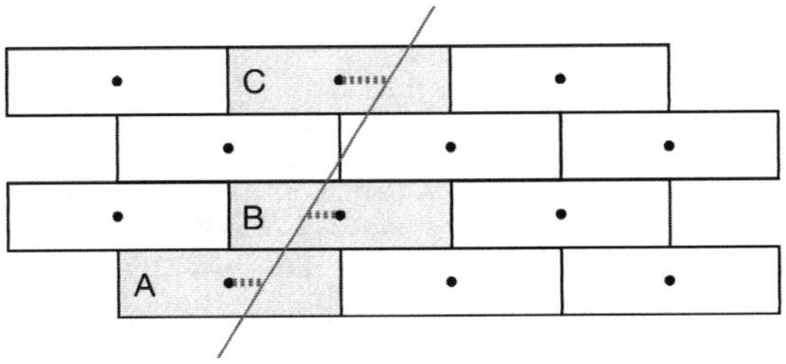

Figure 3. Example of a Mean-Timer equation for one particular triplet of wires and left-right crossing hypothesis (A is right, B is left and C is right), leading to $\mathcal{L}(T_{\text{max}}, T_{\text{A}}, T_{\text{B}}, T_{\text{C}}) = 2T_{\text{A}} + 3T_{\text{B}} + T_{\text{C}} - 2T_{\text{max}} = 0$ and $t_{\text{p}} = (2t_{\text{A}} + t_{\text{B}} + t_{\text{C}} - 2T_{\text{max}})/6$.

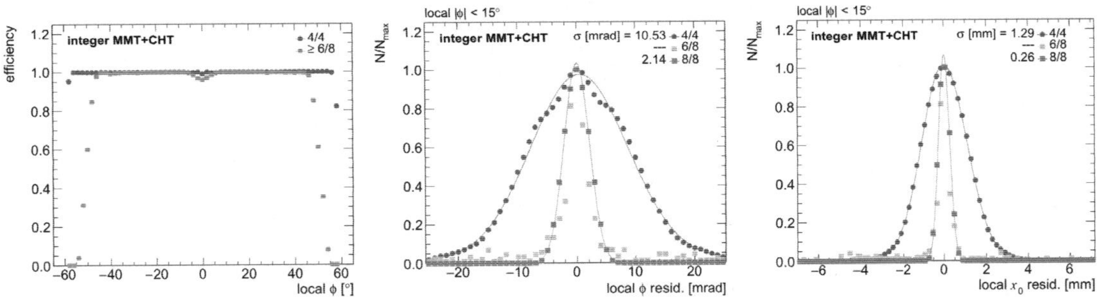

Figure 4. MMT+CHT algorithm efficiency for finding track segments with 4 aligned reconstructed hits in one super-layer (4/4) and with at least 6 aligned reconstructed hits in both super-layers ($\geqslant 6/8$), shown as a function of the local track angle ϕ (left). Resolution of the reconstructed local track angle ϕ (centre) and intercept x_0 (right) with 4 aligned reconstructed hits in one super-layer (4/4), with 6 aligned reconstructed hits in both super-layers (6/8), and with 8 aligned reconstructed hits in both super-layers (8/8), for tracks with local angle $|\phi| < 15°$. The resolution is expressed as the square root of the variance of a weighted sum of two normal functions used to fit the distribution of rhe residuals. Inclusive results are shown.

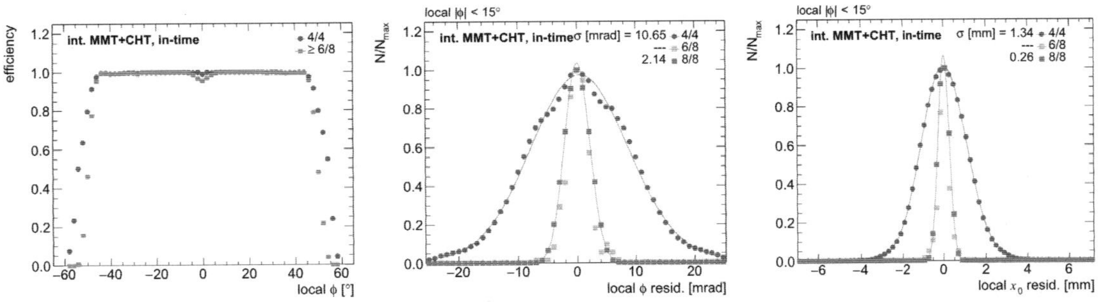

Figure 5. Same as in Figure 4, exclusively for in-time muon candidates, i.e. featuring a correct t_{p} identification.

5 Future perspective and concluding remarks

The design of a HT-based algorithm to be implemented in FPGA targeting the replacement of the CMS DT TPG upgrade for the HL-LHC is currently ongoing and promising. An extensive study of the main parameters of the algorithm has been completed and the resulting fully-integer simulation matches the performance of the floating point one. The BX identification based on Mean-Timer techniques is providing correct results, and there is still some margin for improvements. Also, the BX identification does not add any degradation in terms of resolution. This study will benefit from the evaluation of the results of the first RTL synthesis which will hint to bottlenecks and will suggest improvements. Other valuable input could be available from the inclusion of real-life effects as well as from running the algorithm on CMS collision data.

References

[1] The CMS Collaboration, JINST **3**, S08004 (2008)
[2] N. Pozzobon et al., Nucl. Inst. & Meth. A **824**, 346 (2016)
[3] P. Arce et al., Nucl. Inst. & Meth. A **534**, 441 (2004)
[4] S. Betkhe et al., Nucl. Inst. & Meth. A **416**, 243 (1998)
[5] N. Pozzobon, F. Montecassiano, P. Zotto, Nucl. Inst. & Meth. A **834**, 81 (2016)
[6] P.V.C. Hough, U.S. patent **3,069,654** (1962)
[7] J. Illingworth, J. Kittler, Comp. Vis. Graph. Image Proc. **44**, 87 (1988)
[8] P. Mukhopadhyay, B.B. Chaudhuri, Pattern Recognition **48**, 993 (2015)
[9] M.D. Giorgi et al., Nucl. Inst. & Meth. A **438**, 302 (1999)
[10] F. Gasparini et al., Nucl. Inst. & Meth. A **336**, 91 (1993)

DOI: 10.1051/epjconf/201612700013

Customization of the GENFIT2 Fitting Package in $\overline{\mathrm{P}}$ANDA

Elisabetta **Prencipe**[1] on behalf of the $\overline{\mathrm{P}}$ANDA Collaboration[a]

[1] *Forschungszentrum Jülich, Germany*

Abstract. The availability of a cooled antiproton beam in the energy range from 2.0 to 5.5 GeV will allow the $\overline{\mathrm{P}}$ANDA experiment, planned for operation in FAIR (Darmstadt, Germany), to perform a wide nuclear and particle physics program. $\overline{\mathrm{P}}$ANDA is the only experiment worldwide in that energy regime that combines a solenoidal magnetic field ($B = 2\,\mathrm{T}$) and a dipole field (maximum bending power equal to 2 Tm) in a fixed-target experiment. The offline tracking algorithm is developed within the PandaRoot framework, which is part of the FairRoot project. Here the GENFIT2 package is presented. The tool contains an implementation of the Kalman filter and the deterministic annealing filter for charged particle track reconstruction, together with a Runge-Kutta track representation. It also interfaces with Millipede II and RAVE, for alignment and vertex reconstruction, respectively. The performance of this track-fitting package for the $\overline{\mathrm{P}}$ANDA experiment is shown for the first time, within the PandaRoot framework. For those channels where a good low momentum tracking is required, i.e. $p_{\mathrm{T}} < 350\,\mathrm{MeV}/c$, an improvement by a factor of about two is shown.

1 Introduction

The GENFIT2 package [1] is a standalone tool that provides an extensible modular open-source framework, performing track fitting and other related tasks. It is available via *GitHub*.

Tracking is the core of physics analyses. All particle physics experiments contain different types of tracking detectors, to collect position information and extract from that signal the direction and the momentum of the particles. The tracking detectors must therefore be able to measure with good precision where the particle passed. When tracking is performed inside a magnetic field, it is possible to estimate the momentum of the particle by the trajectory curvature. For local track fitters, such as the Kalman filter, the additional knowledge of the particle mass allows to fully reconstruct the track, and infer the most accurate value of its momentum. Collections of measured detector signals (hits) must be associated to a track candidate, which is then fitted. After the fitting procedure, tracks are then refined by adding missing hits, and gathering the information from the different tracking subdetectors. Trajectories pointing to the same origin must be combined through vertex finding and fitting. All these tasks can be performed by the new GENFIT2 package, given a suitable collection of measurements and the description of the detector geometry.

GENFIT2 extends and improves the work previously performed by GENFIT [2], the external tracking tool available in PandaRoot releases (as part of PandaRoot [3] since 2007, and as independent

[a]e-mail: e.prencipe@fz-juelich.de

open source code since 2009). PandaRoot is based on FairRoot [4], and it is the official software framework for the Monte-Carlo (MC) simulations, event reconstruction and analysis procedures of the $\overline{\text{P}}$ANDA experiment. Several other experiments right now make use of GENFIT and/or GENFIT2; among these are Belle II, FOPI, SHIP, and BGO-OD.

In the following we will not describe the $\overline{\text{P}}$ANDA detector in detail, since it is extensively done elsewhere [5]. Here we mention only the main tracking detector components, as this report intends only to briefly document the implementation of the GENFIT2 classes inside PandaRoot, and its main features and performances therein. $\overline{\text{P}}$ANDA is a fixed-target experiment, with almost full coverage, including different kind of gaseous and solid state detectors. Its tracking system consists of a high performance silicon vertex detector (MVD), a straw-tube central tracker (STT), a gas-electron multiplier (GEM) tracker, and a forward tracking system (FTS).

The present documentation is based mostly on Refs. [1], [2] and [6], and shows additional plots based on the PandaRoot performance only. These have been obtained using the PandaRoot trunk-revision 28747, where GENFIT and GENFIT2 are both available.

The paper is organized as follows: a general introduction is given; in Section 2 the documentation of the GENFIT2 classes is provided, with a general overview of the structure of this modular code, and what was used then inside PandaRoot; Section 3 is dedicated to the GENFIT2 customization inside PandaRoot; Section 4 shows the visualization of a track in the $\overline{\text{P}}$ANDA event display; the final Section 5 summarizes the goals and achievements of this preliminary work.

2 Structure of the code

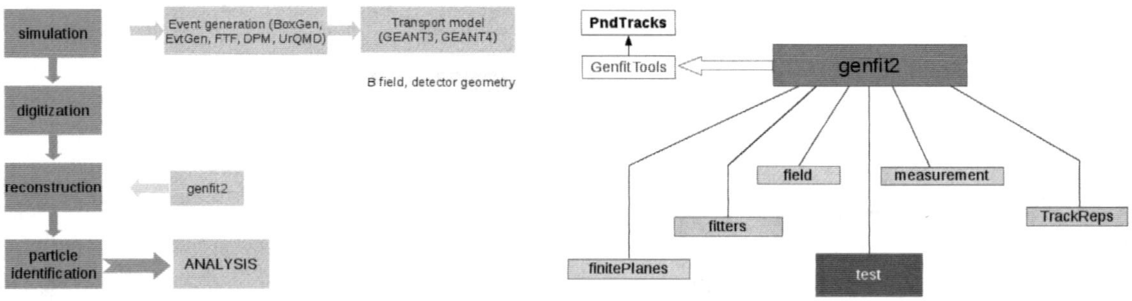

Figure 1. General structure of PandaRoot (left); customization of GENFIT2 in PandaRoot (right). RAVE and the GENFIT2 event display are not yet interfaced with PandaRoot. The PandaRoot package *GenfitTools* works as an interface between the GENFIT (GENFIT2) classes and PandaRoot: Track objects (GENFIT or GENFIT2 tracks) are transformed into PndTrack objects ($\overline{\text{P}}$ANDA tracks).

The GENFIT2 package is a C++ object-oriented modular tool. In Figure 1 the PandaRoot general structure and the GENFIT2 tool classes interfaced to it are shown. Event generation and analysis in PandaRoot is performed in four main steps:

1. Simulation. Events can be produced using different MC generators, that rely on different models to generate the physics process that the user wants to investigate. Several MC event generators are available in PandaRoot: *EvtGen*, Dual-Parton model (*DPM*), Fritiof model (*FTF*),

the Ultrarelativistic Quantum Molecular Dynamics model (*UrQMD*), a single particle generator (*BoxGen*), etc. The output of this process consists of two ROOT files: one containing the simulation parameters, and one the main information about MC tracks and points generated in the $\overline{\text{P}}$ANDA subdetectors: track number, event number, x-y-z coordinates, x-y-z momentum coordinates, time, path length, etc. The detector simulation is performed either by GEANT3 or by GEANT4, according to the choice by the user executing the simulation macro.

2. Digitization. The digitization process runs after the simulation. It simulates the detector response to the passage of particles. At this stage a ROOT file is produced that saves the main information about hits in the $\overline{\text{P}}$ANDA subdetectors.

3. Reconstruction. This is the step where GENFIT (or GENFIT2) is invoked, and where tracks are fitted using the GENFIT (or GENFIT2) methods. The output ROOT file stores all information related to the track combination from different subdetectors: time stamp, time stamp error, space and momentum coordinates of the first and last hit, χ^2 of the track fit, number of degrees-of-freedom NDof, fit quality flags, etc.

4. Particle Identification (PID). This last step runs after the reconstruction process. The PID technique uses a likelihood method, combining variables that help in separating tracks with different mass hypotheses. All variables involved in the PID process can be checked in the output file: $\mathrm{d}E/\mathrm{d}x$, Cherenkov angle, electromagnetic calorimeter EMC lateral moment, Zernike moments, etc.

After these four steps, the user can run the analysis macro, with different settings, depending on the specific simulated decay mode. The selection of events is done in this last step.

2.1 Track fitting in GENFIT2

In GENFIT2, track fitting is based on the following three components:

- measurements;
- track representations;
- fitting algorithms.

The measured coordinates and their covariance matrix are contained in objects of class Measurements, as described extensively in Ref. [1]. The GENFIT2 package includes predefined measurement classes for various detector types. For planar detectors, the detector plane is given by the detector geometry, whereas for WireMeasurements and SpacepointMeasurements virtual detector planes are built.

A track representation consists of a five-dimensional state vector that parameterizes the track and the corresponding covariance matrix. It also provides the possibility to extrapolate the parameters and covariance matrix from one point of the detector to another. The GENFIT2 tool implements a track representation based on a Runge-Kutta extrapolator (RKTrackRep, see Section 2.2). Fitting algorithms use the measurements and track representations to calculate fit results.

All information related to a track is kept in the Track object, which contains both the measurement and track representation objects. The Track holds a sequence of TrackPoints, which contain the Measurements.

GENFIT2 can work with five different track representations, one for each mass hypothesis: electron, muon, pion, kaon, and proton. The one that best fits to the measurement must be selected as the cardinal track representation. The selection can be done either by the user or by GENFIT2 itself. The

Track class also stores general information (number of iterations used by the fit, etc.) and fit properties (χ^2, NDof, etc.).

To summarize, the GENFIT2 class called MeasurementFactory can build a Track out of a list of hits, an object of TrackCand. This Track can then be processed by the various fitting algorithms, described in the next Sections. The user can eventually decide which Track data to keep inside the object, depending on the needs.

2.2 Runge-Kutta track representation

The Runge-Kutta track representation (RKTrackRep) is based on a Runge-Kutta extrapolator from GEANT3 [7]. It uses five parameters, namely $q/p, u', v', u, v$, and their covariance matrix. GENFIT2 can actually use different coordinate systems: 6D-coordinates (x, y, z, p_x, p_y, p_z), and the charge; local coordinates $(q/p, u', v', u, v)$; and global coordinates $(x, y, z, T, q/p)$. These variables are defined as the space coordinates x, y, z; the momentum coordinates p_x, p_y, p_z; the charge q; the absolute value of the momentum p; the direction unit vector T; the local coordinates on the plane u and v; the direction cosines relative to the plane reference frame u' and v'..

The Runge-Kutta calculations are used as track-following routines, to perform track propagation inside the detector geometry, by means of a sequence of steps. In each step, the physical processes that might occur in those particular experimental conditions are taken into account.

An abstract interface class interacts with the detector geometry, either via implementations using the ROOT class TGeoManager [8] or via the GEANT4 class G4Navigator [9]. During the fitting procedure, several effects are computed, that are related to material properties: the mean energy loss and its straggling for charged particles according to the Bethe Bloch formula and the Landau distribution, respectively; the multiple scattering (using the Highland-Lynch-Dahl formula [10, 11]), where the full noise matrix is calculated; the soft Bremsstrahlung energy loss and its straggling for electrons and positrons.

Step sizes for the Runge-Kutta propagation are evaluated inside the RKTrackRep, taking into account field inhomogeneities, track curvature and the above mentioned material effects. RKTrackRep provides different methods to find the point of closest approach (POCA) of the track to non-planar measurements. These are used to construct virtual detector planes, on which the measurement and its covariance matrix are projected.

2.3 Fitter implementation

GENFIT2 provides additional features compared to the old GENFIT package. Four different track-fitting algorithms are available:

- two Kalman filters, one which linearizes the transport around the state predictions, the other which linearizes around a reference track. This is an original feature of the GENFIT2 fitting tool;
- a deterministic annealing filter (DAF);
- a general broken lines fitter.

Every TrackPoint can store several Measurements, and the DAF can assign them weights. In GENFIT2, WireMeasurements and SpacepointMeasurements are implemented, depending on the detector and the user needs. The WireMeasurements class can also produce MeasurementOnPlane objects, representing the passage of the particle on both side of the wire. The GENFIT2 tool provides several ways to handle multiple measurements:

- the weighted average of the individual Measurements is calculated, and this is used by the DAF;
- the Measurements object closest to the state prediction or reference state is selected;
- in case the TrackPoint is a wire measurement, the side that is closest to the prediction or reference is selected; otherwise, the average is selected.

The WireMeasurements class has been tested in Belle II. It has been demonstrated that the best option to run the Kalman filter and obtain reasonable results is to select the side closest to the state prediction. Several iterations of the fit are run and in order to decide whether the fit has converged or not, the difference of the p-value between one iteration and the previous one is computed. When a predefined number of iterations is reached, the procedure stops. Also, the relative change in χ^2 between iterations is checked: if it is larger than 20%, then the fit continues. All parameters, such as the number of iterations and the threshold for the p-value, are user-adjustable.

2.4 Kalman filter implementation

The Kalman filter is a recursive version of a least-square fit. It iterates three steps: extrapolation, filter, and smoothing. It is based on a set of mathematical equations, that uses a series of measurements observed over time, containing "noise" (random effects) and other inaccuracies, and produces estimates of unobservable variables. It is a versatile tool with applications in many fields, and has been successfully used in high-energy physics for many years.

The Kalman filter implemented in GENFIT2 is adapted from the formulas given in Ref. [12], where the linearization is made around the state predictions. Additionally, a square-root formalism, adapted from Ref. [13], provides greater numerical stability at the cost of a larger execution time.

In general, state predictions start from a preliminary estimate of the track parameters. They might fall far from the real trajectories. As a consequence, using them as the linearization point is not an optimal solution; material and magnetic field lookups are also not done at the proper places. A general common solution is to linearize around reference states [12]: the start parameters to all TrackPoint objects are calculated from those. The smoothed states from the previous iterations are used as linearization points. However, for very small changes, reference states are not saving computing time. In addition to the convergence criteria detailed above, the fit is declared as converged if none of the reference states has been updated since the previous iteration.

2.5 Deterministic annealing filter

The deterministic annealing filter (DAF) [14] is a powerful tool used for the rejection of outlying measurements. It is a robust Kalman filter that makes use of a weighting procedure between iterations based on the measurement residuals to determine the proper weights. The user can select which of the two Kalman filter implementations should be used and can specify the annealing scheme.

The DAF is suited to resolve the left-right ambiguities of wire measurements. Both left and right measurements must be initialized. A standard solution is to start with the same weight, equal to 0.5. The wire positions are taken as measurements in the first iteration, and their covariance is twice the mean of the individual covariances. In this respect, all the wire positions have the same covariance value, independently on how far from the real trajectory they are. In this way the systematic indetermination affecting the covariance estimates can bias the fit result.

In GENFIT2, the fit is improved with a new method for the weight initialization: points with smaller drift radii get larger weights. In these cases it is safe to assume that the wire position is closer to the point where the particle actually passed; opposite applies to points with larger drift radii.

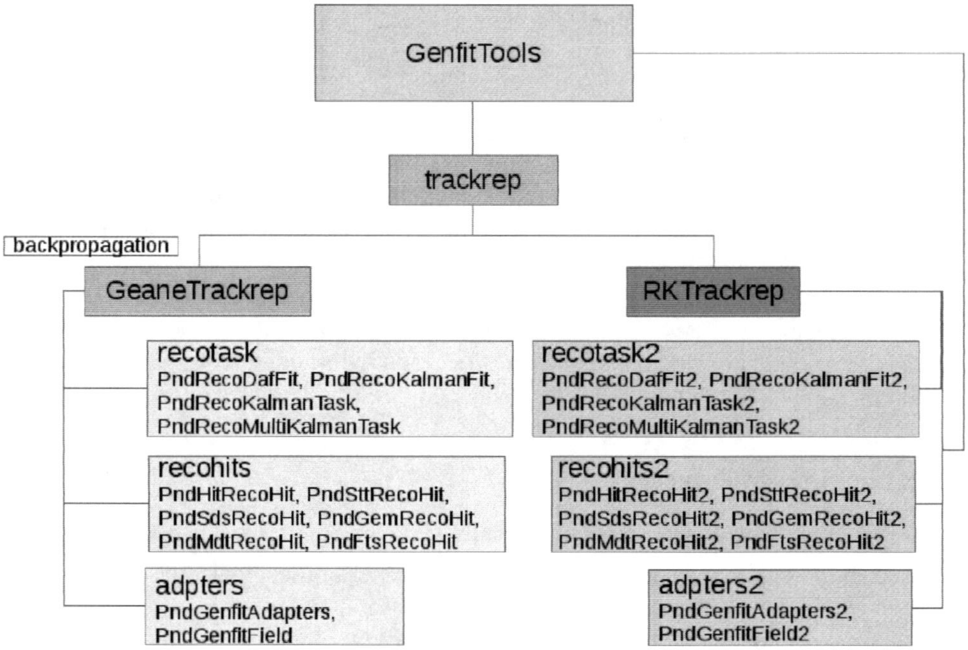

Figure 2. Structure of the *GenfitTools* package in the PandaRoot trunk-revision 28747.

2.6 Generalized broken lines fitter

The generalized broken lines method of track fitting [15] was implemented specifically for the purpose of alignment with Millipede II. It is equivalent to the Kalman filter (with thin scatterers instead of continuous materials), but fits tracks globally in one step, providing a natural interface to the Millipede II software. For alignment purposes, GENFIT2 also provides a set of interfaces for alignment parameters and derivatives which can be implemented by the detector classes.

2.7 Vertex reconstruction with *GFRave*

GFRave, an interface to the vertex-fitting framework RAVE [16], has been implemented in GENFIT2. RAVE is a detector-independent toolkit developed for the CMS experiment [17]. It deals with both finding (pattern recognition) and fitting (statistical estimation) of vertices. It takes a set of reconstructed particle tracks as an input, and produces reconstructed interaction vertices as an output. *GFRave* will allow precise and fast vertex reconstruction, although it is not yet interfaced with PandaRoot.

3 Performance in PandaRoot

An interface was built between GENFIT2 and PandaRoot, known as the package *GenfitTools*. The basic structure of *GenfitTools* is the same of the old GENFIT package, but with a few modifications, due to the new classes introduced now in GENFIT2 (see Figure 2). The new *GenfitTools* structure was created to allow the use of both GENFIT and GENFIT2 as fitting tool, and then to compare results.

GenfitTools includes several classes: adapters and adapters2; recohits and recohits2; recotasks and recotasks2; and trackrep.

DOI: 10.1051/epjconf/201612700013

The classes adapters, recohits, and recotasks refer to the GENFIT interface to the PandaRoot classes, and the classes adapters2, recohits2, recotasks2 refer to the GENFIT2 interface to the PandaRoot classes. The class recohits takes care of the hit reconstruction in different detectors; in the class trackrep the information about track representation is accessible; the class recotasks contains the functions needed to reconstruct tracks; the class adapters (adapters2) is the PandaRoot class where the transformation of a GENFIT (GENFIT2) track to the PndTrack is performed, and vice versa. The user should use objects of the PndTrack only, so whenever a GENFIT2 update is provided, information is properly stored in the PandaRoot objects and there is no need to change other PandaRoot packages.

In order to evaluate the capabilities of the GENFIT2 and the new *GenfitTools* package, first tests have been performed using the $\overline{\text{P}}$ANDA package *BoxGen*.

Five particle hypotheses have been tested: electrons, muons, pions, kaons, and protons. We have generated separately single particle events for each type, using different values of p_T. We tested particles at a fixed polar angle $\theta = 60°$. We also checked the GENFIT2 performance in PandaRoot by changing the beam momentum, to evaluate the resolution and pull distributions of the tracking parameters.

The magnetic field map used in these tests was a realistic one, as shown in Figure 3. This is obtained by converting TOSCA [18] calculations into PandaRoot format.

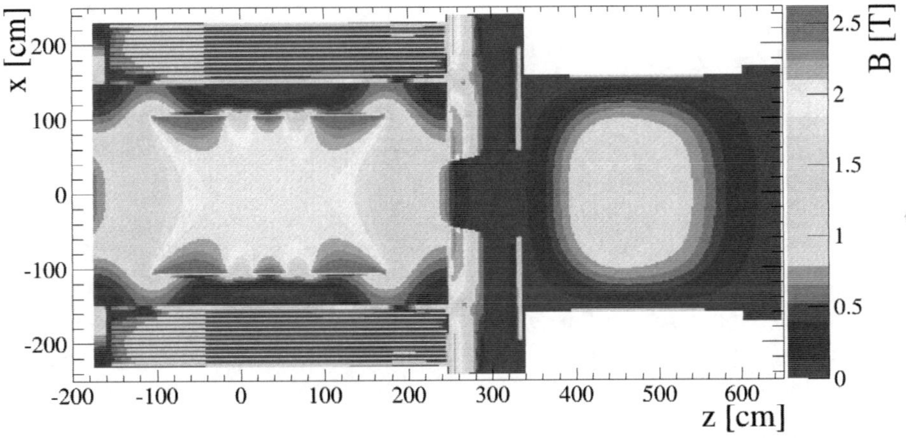

Figure 3. Magnetic field scheme in $\overline{\text{P}}$ANDA. In the central part of the detector, the B field is constant and equal to 2 T, and in the dipole area the maximum bending power is 2 Tm.

Figure 4 shows the reconstruction efficiency of single pions, simulated with the MC *BoxGen* package, as a function of p_T. A comparison between GENFIT2 (blue points), and GENFIT (red points) is given: a clear improvement for low momentum tracks is observed with GENFIT2. In this test, 25000 single particle events were generated, for each p_T value. Efficiency is evaluated as the ratio between the number reconstructed events, selected in a 3σ window around the momentum peak, and the number of generated events. The resolution $\Delta p/p$ at 1 GeV/c is about 1.8%, and the ϕ angle resolution is about 2.5 mrad. The efficiency shown in the plot takes into account the detector acceptance and a realistic simulation of the detector geometry, digitization and track reconstruction.

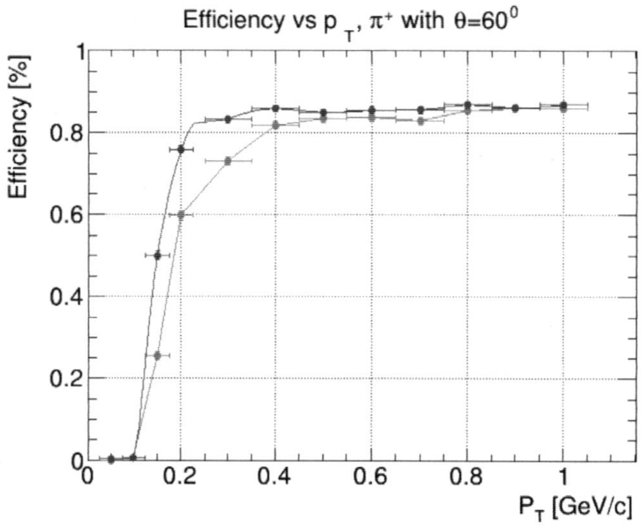

Figure 4. Reconstruction efficiency of charged pions emitted at $\theta = 60°$, as a function of p_{T}. Single particle events were produced by the MC package *BoxGen*. The blue points show the performance of the GENFIT2 package in the PandaRoot trunk-revision 28747, compared to GENFIT (red points). A clear improvement is observed for $p_{\mathrm{T}} < 350\,\mathrm{MeV}/c$ when using GENFIT2.

A transverse momentum cut larger than $50\,\mathrm{MeV}/c$ is applied, because the Pattern Recognition tool requires at least in the STT. The minimum radius of the STT volume is 16 cm, so the curvature radius of the charged particle must be larger than 8 cm. This requirement leads to a minimum p_{T} value of $48\,\mathrm{MeV}/c$ for every tracked particle in the $\overline{\mathrm{P}}$ANDA detector.

Figure 5 shows the Gaussian fit to the distributions of the standardized residuals of the track parameters d_0 and z_0, for $p_{\mathrm{T}} = 1.0\,\mathrm{GeV}/c$. These two variables are defined as the distance of the point of closest approach to the origin (0,0,0) in the x-y plane and z axis, respectively. The standardized residual is defined as the difference of the generated values and the reconstructed value of the variable, divided by its error. The width of the Gaussian fit to the residuals is close to one, which is expected.

4 Visualization

The GENFIT2 package also provides a 3D event display. Not only the fitted tracks can be shown (also "live" while performing the fit), but the geometry, all the hits (measured, extrapolated, and smoothed) and their reference planes can be drawn as well.

In PandaRoot, the package called *PndEventDisplay* provides similar features, derived from the ROOT EVE package [19]. Therefore, the GENFIT2 event display is not interfaced with PandaRoot. In Figure 6, an example of events observed with *PndEventDisplay* is shown.

5 Conclusions

We have demonstrated that GENFIT2 improves low momentum tracking for pions. Similar studies have ben performed using other particle types: electron, muon, kaon, and proton. The new track representation available in GENFIT2 works better than the previous one used in PandaRoot. Examples

DOI: 10.1051/epjconf/201612700013

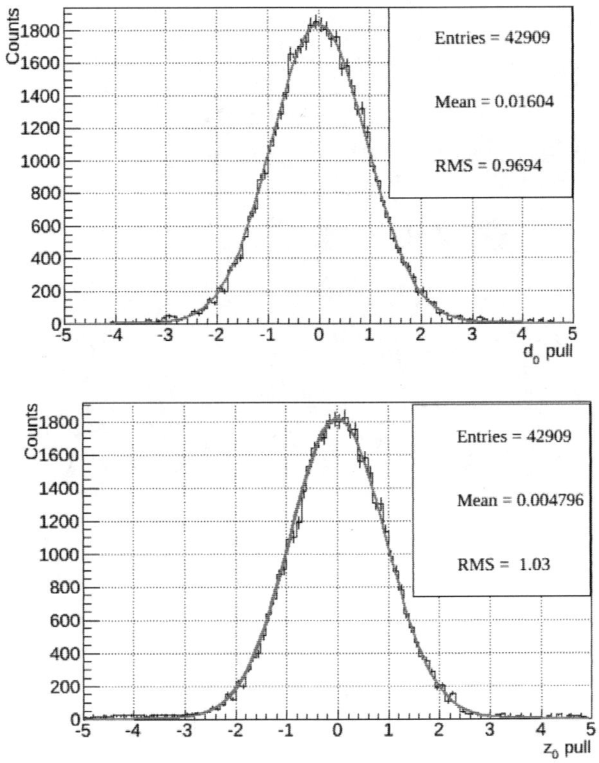

Figure 5. Distribution of the standardized residuals of d_0 (top) and z_0 (bottom), and the Gaussian fit to the distributions, for $p_T = 1.0\,\text{GeV}/c$.

with electron reconstruction were shown during the workshop in Ref. [20]. The future plan is to use GENFIT2 as a part of a PandaRoot release for MC simulations. Tests are ongoing on some benchmark channels of the $\overline{\text{P}}$ANDA physics program, more sensitive to the reconstruction of low momentum tracks, for instance channels involving the reconstruction of D_s mesons. Preliminary results are promising.

Acknowledgments

Special thanks go to Lia Lavezzi for the extremely valuable work on tracking tools performed inside the PandaRoot framework; and to Johannes Rauch and Tobias Schlüter, main authors of the GENFIT2 package [1], by which most of Section 2 of this report is inspired, for kind clarifications and fruitful discussions, and for their hard work in developing this track fitting tool.

References

[1] J. Rauch, T. Schlüter, J. Phys. Conf. Ser. **608**, 12042 (2015)
[2] C. Höppner et al., Nucl. Inst. & Meth. A **620**, 518 (2010)
[3] D. Bertini et al., J. Phys. Conf. Ser. **119**, 32011 (2008)

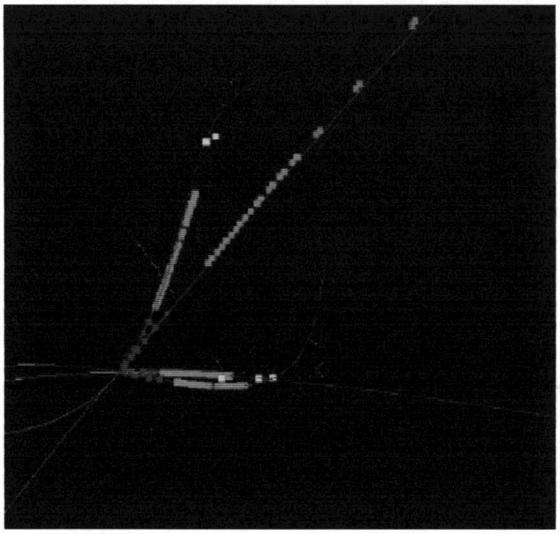

Figure 6. Example of an event in the $\overline{\mathrm{P}}$ANDA detector. The different colors refer to hits and tracks in the different sub-detectors: blue for MVD, magenta for STT, red for GEM, cyan for SciTil (scintillator tile hodoscope).

[4] S. Spataro, J. Phys. Conf. Ser. **396**, 22048 (2012)
[5] W. Erni et al. (PANDA collaboration) (2009), `arXiv:0903.3905`
[6] I. Gaponenko et al., Comp. Phys. Comm. **140**, 406 (2000)
[7] R. Brun et al., *GEANT: detector description and simulation tool* (CERN, Geneva, 1994), `http://cds.cern.ch/record/1082634`
[8] R. Brun, F. Rademakers, Nucl. Inst. & Meth. A **389**, 81 (1997)
[9] S. Agostinelli et al. (GEANT4), Nucl. Inst. & Meth. A **506**, 250 (2003)
[10] V.L. Highland, Nucl. Inst. & Meth. **129**, 497 (1975)
[11] G.R. Lynch, O.I. Dahl, Nucl. Inst. & Meth. B **58**, 6 (1991)
[12] R. Frühwirth et al., *Data Analysis Techniques for High-Energy Physics* (Cambridge University Press, Cambridge, UK, 2002)
[13] B.D.O. Anderson, J.B. Moore, *Optimal Filtering* (Dover Publications, Mineola, NY, 1979)
[14] R. Frühwirth, A. Strandlie, Comp. Phys. Comm. **120**, 197 (1999)
[15] C. Kleinwort, Nucl. Inst. & Meth. A **673**, 107 (2012)
[16] W. Waltenberger, IEEE Trans. Nucl. Sci. **58**, 434 (2011)
[17] S. Chatrchyan et al. (CMS collaboration), JINST **3**, 361 (2008)
[18] Vector Fields Limited, *The TOSCA Reference Manual*, Vol. VF068814 (24 Bankside, Kidlington, Oxford, OX5 1JE, England, 1988)
[19] M. Tadel, J. Phys. Conf. Ser. **219**, 42055 (2010)
[20] E. Prencipe, *Customization of GENFIT2 fitting package for PANDA*, Talk at "Connecting the Dots 2016" workshop (2016)

EPJ Web of Conferences **127**, 00014 (2016) DOI: 10.1051/epjconf/201612700014
Connecting The Dots 2016

Online Reconstruction and Calibration with Feedback Loop in the ALICE High Level Trigger

David **Rohr**[1,a], Ruben **Shahoyan**[2], Chiara **Zampolli**[2,3], Mikolaj **Krzewicki**[1,4], Jens **Wiechula**[4], Sergey **Gorbunov**[1,4], Alex **Chauvin**[5], Kai **Schweda**[6], and Volker **Lindenstruth**[1,4] for the ALICE Collaboration

[1] *Frankfurt Institute for Advanced Studies, Frankfurt am Main, Germany*
[2] *CERN, Geneva, Switzerland*
[3] *INFN Sezione Bologna, Italy*
[4] *Goethe University, Frankfurt, Germany*
[5] *Technical University, Munich, Germany*
[6] *University of Heidelberg, Germany*

Abstract. ALICE (A Large Heavy Ion Experiment) is one of the four large scale experiments at the Large Hadron Collider (LHC) at CERN. The High Level Trigger (HLT) is an online computing farm, which reconstructs events recorded by the ALICE detector in real-time. The most computing-intensive task is the reconstruction of the particle trajectories. The main tracking devices in ALICE are the Time Projection Chamber (TPC) and the Inner Tracking System (ITS). The HLT uses a fast GPU-accelerated algorithm for the TPC tracking based on the Cellular Automaton principle and the Kalman filter. ALICE employs gaseous subdetectors which are sensitive to environmental conditions such as ambient pressure and temperature and the TPC is one of these. A precise reconstruction of particle trajectories requires the calibration of these detectors. As our first topic, we present some recent optimizations to our GPU-based TPC tracking using the new GPU models we employ for the ongoing and upcoming data taking period at LHC. We also show our new approach to fast ITS standalone tracking. As our second topic, we present improvements to the HLT for facilitating online reconstruction including a new flat data model and a new data flow chain. The calibration output is fed back to the reconstruction components of the HLT via a feedback loop. We conclude with an analysis of a first online calibration test under real conditions during the Pb-Pb run in November 2015, which was based on these new features.

1 Introduction

1.1 The ALICE Detector

ALICE (A Large Heavy Ion Experiment) is one of the four major experiments at the Large Hadron Collider (LHC) at CERN in Geneva [1]. While the other large experiments focus mainly on proton-proton collisions, the main purpose of ALICE is to study heavy ion collisions. This enables the

[a] e-mail: drohr@jwdt.org

investigation of matter under extreme conditions of high temperature and pressure. In ion physics mode, the LHC collides lead nuclei at an interaction rate of around 8 kHz. ALICE employs several detectors to measure particle trajectories, energy deposition, and to identify particles.

1.2 The High Level Trigger (HLT)

The High Level Trigger (HLT) [2] is an online computing farm consisting of around 200 computing nodes for the online processing of the collisions recorded by ALICE. In contrast to the subsequent offline physics analysis, the HLT performs the first processing and analysis in real time. This involves the reconstruction of the events, calibration of the detectors, data compression to reduce the amount of data stored to tape, online quality assurance (QA), as well as triggering the readout or tagging of physically interesting events. The HLT receives the data from the experiment via several hundred optical links. Inside the HLT, independent processing components perform individual steps of the reconstruction and processing. A custom data transport framework transfers the data between the processing components on the same servers via shared memory or on different servers via an Infiniband network. The maximum possible data input rate over the detector links is above 60 GB/s, while in normal operation the HLT receives up to 30 GB/s of recorded data. The HLT is capable of full real time event reconstruction of the data recorded by ALICE. The computationally most intensive step is the reconstruction of particle trajectories, also called tracking.

1.3 Online reconstruction and online calibration

Several of the ALICE subdetectors are sensitive to environmental conditions such as ambient pressure and temperature. Precise reconstruction of particle trajectories requires the calibration of these detectors. Since the environmental conditions change during data taking, calibration must be performed regularly, as a single calibration step at the beginning of a run is insufficient. Performing the detector calibration (or a part thereof) online in the HLT has several advantages:

- Making the calibration result available to the online reconstruction in the HLT can significantly improve the quality of online reconstruction.
- Performing the calibration while the data is recorded allows an immediate and better QA already during data taking.
- Online calibration can render certain offline calibration steps redundant, thereby reducing the computational load during offline reconstruction.
- Looking ahead to future experiments such as ALICE in LHC Run 3 or FAIR at GSI [3], data compression will rely on reconstruction, making online calibration a necessity.

2 Approach to online calibration

On one hand, the ALICE calibration is based on reconstruction results like particle trajectories; on the other hand, the calibration results should be used to improve the reconstruction. This imposes a cyclic dependency between reconstruction and calibration. On top of that, calibration involves long running tasks that can last many seconds or even minutes. It is therefore impossible to apply the calibration result to the reconstruction of the events that are used for the calibration. This would require caching of data for a considerable amount of time which is not possible in the HLT.

Instead, the HLT employs a different approach to online calibration. The ambient conditions that affect the calibration are stable over a certain time period. Even in case of a sudden total weather change, pressure and temperature will change smoothly, so that the calibration created for a certain point in time is valid for a certain period. In the following, we will discuss the TPC drift velocity as an example. In this case, the calibration is assumed to be valid for the following 15 minutes. The calibration is performed in multiple consecutive intervals organized in a pipeline with three steps:

- **Step A**: Incoming data is reconstructed using the last valid calibration (or the default calibration at the very beginning). Based upon the reconstruction, new calibration is computed. This is performed as long as needed such that sufficiently many events are processed to produce valid calibration.
- **Step B**: The calibration result is propagated back to the beginning of the reconstruction chain such that it can be applied to the reconstruction. Necessary postprocessing steps are performed as well such as the preparation of a new TPC cluster transformation map[1].
- **Step C**: The new calibration is now used for the reconstruction as long as it is valid, or until a new calibration is available.

Figure 1 illustrates the online calibration scheme. The steps are processed in a pipeline, such that while the calibration result is fed back (Step B, brown) and then used in the reconstruction (Step C, purple), a new calibration is computed in parallel (Step A, blue). Accordingly, the reconstruction is not calibrated for events recorded at the very beginning of a run before Step B finishes. The final reconstruction objects stored at the end of the process for offline use are valid for all events, also those at the beginning. This scheme is feasible as long as the total time of steps A, B, and C is below the stability interval, e. g. 15 minutes in case of the TPC drift time calibration.

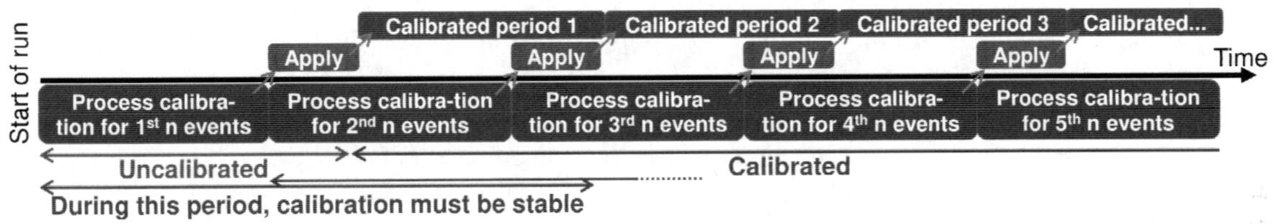

Figure 1. Illustration of the online calibration scheme. The blue boxes are the intervals where the calibration data is aggregated. Afterwards there is a short delay to prepare new TPC transformation maps based on the calibration and distribute them in the cluster (brown box). Finally, the HLT reconstruction uses the calibration as long as it is stable (purple box).

Naturally, a single instance of the calibration software running on a single processor cannot compute the calibration objects in time. The HLT runs the calibration component on 172 of its 180 computing nodes with 3 instances of the calibration per computing node, for a total of 516 calibration processes running in parallel. These instances process the incoming events in a round-robin fashion and regularly send their calibration data to a single calibration merger process running on a dedicated calibration computing node. This calibration node merges all the calibration data and creates the final calibration objects. The objects are stored for offline use and shipped back in parallel to all computing nodes by the feedback loop, such that the computing nodes can perform the reconstruction based on the new calibration. Figure 2 illustrates the process.

[1]The TPC cluster transformation map is needed to convert the raw TPC clusters from TPC pad row, TPC pad, and time coordinates into x, y, and z coordinates taking into account the calibration.

DOI: 10.1051/epjconf/201612700014

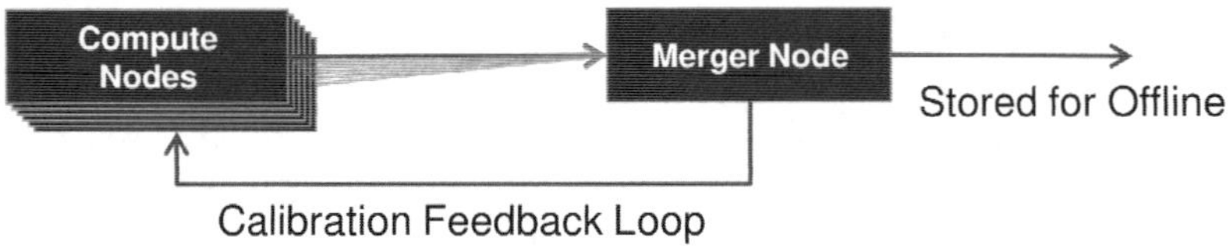

Figure 2. Illustration of the merging of calibration objects.

2.1 Requirements for online calibration

On top of the time constraints from the above approach, online calibration poses a couple of requirements on the reconstruction and data transport framework, which are discussed in the following. This subsection concludes with a list of requirements that needed to be solved to run the TPC drift time calibration in the ALICE HLT. The following sections show how ALICE deals with these challenges.

The custom data transport framework in the HLT can be seen as a directed graph. The fibres from the detector are input nodes in the graph, the network connections to the data acquisition (DAQ) are output nodes, and the remaining nodes in the graph are processing components. The links in the graph define the data flow. All processing components process the events in an event-synchronous way, i. e. they process one event after another in a pipeline. Long-running tasks as needed for the calibration pose a problem, because they stall the pipeline during the processing of the event. All processing components which are upstream of the stalled process become stalled as well as soon as the buffers run full. Another issue is that the HLT processing chain must be loop-free for technical reasons by design, because of the event-synchronous approach. A loop in the graph would impose a cyclic dependency, i. e. two processing components would wait for each other to finish processing the same event.

The TPC drift time calibration matches tracks reconstructed in the TPC to tracks in the ITS. This means the HLT must at least provide track reconstruction for the TPC and the ITS. On top of that, the track reconstruction in these detectors should be standalone, i. e. independent of each other, in order to avoid introducing a bias. Although the ITS delivers significantly less data than the TPC, tracking in the ITS is computationally intensive because of combinatorics, as the ITS sits in the center of ALICE where the track density is the highest. Also, the treatment of scattering in the silicon layers is more complicated than inside the gas volume of the TPC. Finally, ITS tracks have only up to six hits compared to more than 100 in the TPC, which requires a much more robust seeding procedure in order to find all tracks. Therefore, the default HLT method for ITS tracking is to prolong TPC tracks into the ITS and then collect the ITS hits close to the extrapolated tracks. This poses two problems: first, it could introduce a bias to use prolonged TPC tracks for the calibration of the TPC. Second, the prolongation into ITS where the track density is very high requires high precision and works well only if the TPC is calibrated. This leads to a chicken and egg problem.

Standard physics analysis in ALICE is based on the ROOT analysis and statistic software package. The reconstruction creates C++ ROOT objects (see Section 5), which are then used by the calibration tasks. One paradigm in the ALICE HLT for online calibration is to use the same code for the calibration tasks that is used offline. This reduces code duplication and simplifies the verification of the calibration code. Processing components in the HLT are individual processes which cannot exchange C++ objects via pointers because they do not share a common address space. The standard approach is serialization and deserialization of the object which causes significant CPU load. Hence, HLT software should use only flat data structures which can be shipped to other processes directly. This is incompatible with the standard offline reconstruction programming model.

Usually, calibration is performed relative to a default calibration. For instance, the TPC measures the clusters in row, pad, and time coordinates which must be transformed into spatial coordinates in order to run the track reconstruction and then the calibration. This transformation is performed using the initial, default calibration. The calibration uses the transformed clusters, and the exact transformation that was used to obtain the spatial coordinates of the clusters has an impact on the calibration output. In other words, the calibration must run on clusters transformed according to the default calibration, but not on calibrated clusters.

The following list summarizes the requirements for online calibration of the TPC drift in the ALICE HLT:

- Fast reconstruction algorithms for ITS and TPC.
- Independent standalone tracking algorithms for ITS and TPC.
- Support for long-running tasks in the HLT framework.
- Support for loops in the HLT data flow.
- Data structures enabling fast data exchange between processing components.
- The feedback loop must apply the calibration only to the tracking, but not to the calibration component.
- Calibration process and feedback loop must not take longer than the time interval during which the calibration remains stable.

2.2 Framework improvements

Adding the feedback loop directly to the loop-free HLT processing chain is a major architectural change. We wanted to avoid this since the current HLT data transport is thoroughly tested and we preferred incremental changes. In particular, we prefer adding additional processing or communication components over changing the basis of the framework. Considering the general approach to online calibration in the HLT, the calibration does not need to be event-synchronous. The calibration result is not fed back to the reconstruction of the same event, but at some later point in time. This can happen asynchronously. The HLT uses additional source and sink components based on the ZeroMQ data transport library for new communication channels not foreseen in the original framework [4, Subsection 5.3].

The processing rate in the HLT is usually between 1 kHz and 3 kHz. It is impossible to run long-running calibration tasks that last many seconds at this rate due to limited CPU resources. Hence, the calibration task has to run only for a subset of the events, which is totally sufficient to gather enough statistics for the calibration. Still, a long-running task blocks the chain even if it processes only a fraction of the events. The problem is that it stalls the processing of that event for too long such that it affects all the other processes in the chain. This means, even if the average processing time of the events would be short enough because many events are skipped, single events that need much time already block the chain. This deficit of the event-synchronous processing approach is overcome by complementing the HLT with asynchronous processing components, which spawn a subtask in an asynchronous individual process, and then continue the fast synchronous event processing. The result of the subtask is then used as soon as it is available [4, Subsection 5.1]. In order to protect ALICE data taking from fatal errors in the calibration code, the asynchronous processing can optionally happen in a completely isolated operating system process. In this way, a possible memory leak or segmentation violation does not interfere with data taking, but only breaks the processing of the calibration for few events until the process is restarted.

3 Track reconstruction in the TPC

The HLT employs GPU-accelerated track reconstruction for the ALICE TPC that is based on a Cellular Automaton principle to build track seeds, i.e. short track candidates of around 5 to 10 hits. It then uses the Kalman filter for track fitting and track following [5, 6]. The HLT employed 64 NVIDIA Geforce GTX480 GPUs during LHC Run 1. The new HLT farm for LHC Run 2 is now equipped with 180 AMD FirePro S9000 GPUs. One major concern with the original GPU tracker code was that it was based exclusively on the NVIDIA CUDA framework and was thus vendor-dependent. For Run 2, an OpenCL implementation was created, which uses the AMD OpenCL C++ kernel language extensions. The code is written in a generic way, such that the same source code can be used with CUDA, OpenCL, and also for the CPU [4, Section 4]. This gives the greatest possible flexibility for hardware selection and reduces the maintenance effort.

Parallelization is implemented such that during the Cellular Automaton phase, one GPU thread handles one TPC cluster. During the track following, one GPU thread handles one TPC track. This allows for simple and efficient parallel processing, as the threads can operate almost fully independently. Considering the amount of TPC clusters and tracks in a typical Pb-Pb event as well as the number of threads a GPU executes concurrently, this scheme resulted in full GPU utilization at the time it was implemented, e. g. for the GTX480 GPUs during Run 1. Naturally, pp events with much fewer tracks do not use the GPUs efficiently. This is no problem however, as the GPUs are fast enough for pp reconstruction anyway. In the meantime, the number of threads a GPU needs to execute in parallel to achieve full performance has increased significantly. The number of tracks increased only slightly when the LHC moved from $3.5\,\mathrm{TeV}/Z$ to $6.5\,\mathrm{TeV}/Z$ in Run 2. Overall, we see that now even single central Pb-Pb events are unable to load the S9000 GPUs of Run 2 to the full extent. Looking ahead to Run 3, this problem becomes even more severe because new GPUs will feature even more parallelism.

The maximum data rate the TPC can deliver to the HLT is defined by the number of optical links times the link speed. The HLT was tested with data replay at this maximum possible speed and was proven to be able to run the full TPC reconstruction, while still retaining some margin to run reconstruction for other detectors. Thus, development of HLT TPC track finding for Run 2 is complete. Now, the focus lies on testing improvements for the Run 3 online computing facility already in the HLT during Run 2.

As a first step, we increase the parallelism by processing multiple events on one GPU concurrently. While during the implementation of the first GPU tracker version, GPUs could only execute one kernel at a time, modern GPUs can execute many independent kernels, even from different host applications, in parallel. The HLT can run several instances of the tracker on one GPU processing multiple events concurrently, as long as there is enough GPU memory, which in case of Run 2 is sufficient for 3 central Pb-Pb events with pileup. Table 1 shows a first result. Naturally, the wall time for a single event increases, which is not relevant. (For reference, the time between a central Pb-Pb event reaches the HLT input until it leaves the HLT is in average around 3 seconds.) In contrast, the processing throughput increases by 31.8 %. Using all computing nodes at full capacity, the HLT can reconstruct 40 million TPC tracks per second.

Fortunately, the foreseen TPC readout scheme for Run 3 plays into our hands. ALICE plans a TPC upgrade with continuous readout. The online computing facility will no longer process single events but time frames with many overlapping events. This will offer enough parallelism to use the GPUs to the full extent. It is not clear yet whether the GPU memory will be sufficient to process the track finding of an entire time frame at once. One solution is to slice the time frames along the time-beam axis, process the slices individually, and merge the track segments afterwards. The feasibility of this approach is already shown by the track reconstruction in the current HLT, which processes the TPC sectors individually but concurrently, and merges the tracks segments afterward [5].

Another foreseen development for Run 3 is the porting of additional online reconstruction components on the GPU, in particular, because modern GPUs can execute multiple kernels at the same time. Canonical candidates for GPU adaptation are the ones before and after the GPU track finding in the HLT reconstruction chain (see Fig. 7 in the last section). As a first prototype, the final track refit, a sub-step of the TPC track merger and track fit component that merges the track segments reconstructed by the GPU track finding was ported to GPUs. The prototype needs in average 6.8 ms per event for the refit compared to 125.5 ms on a single CPU Core (Intel Nehalem 2.8 GHz, the same events as in Table 1). The bottleneck in this case is the PCI Express transfer, which takes longer than the computation itself. Therefore, the most reasonable approach is to bring multiple successive components of the HLT chain onto the GPU, such that the data need not be transferred back and forth in between.

We plan to implement every new GPU processing component using a generic shared source code for CPU and GPU, in the same way as for the TPC tracking. In addition, the implementations should be flexible enough to be applicable in the new software framework for online computing in Run 3 but also in the current HLT framework. This will allow us to test new developments and benefit from them already in Run 2.

4 Track reconstruction in the ITS

4.1 Scheme of ITS tracking in the HLT

The initial ITS tracking in the HLT which starts from prolonged TPC tracks is unsuited for online calibration. Conversely, a full standalone ITS tracking has to deal with the excessive combinatorics inside ITS and would need too many CPU resources. The HLT thus employs a hybrid approach with two independent ITS tracking branches. The first is the traditional chain with prolonged TPC-ITS tracks. In order to ensure good tracking results, the TPC needs to be calibrated. In parallel, a second chain performs fast standalone ITS tracking that has some limitations. In particular, it is not required to have maximum efficiency, i. e. it does not need to find all tracks. It only needs to find sufficiently many tracks for the ITS–TPC matching in the calibration and for the luminous region estimation (see Section 4.3). The scheme is visualized in Fig. 3. The following subsection describes the ITS standalone tracking in the HLT.

4.2 Fast ITS standalone tracking

The aim of the online ITS standalone reconstruction is to provide a sample of primary ITS tracks sufficient for calibration, without attempting to maximize the track-finding efficiency. Instead, the emphasis is put on the processing speed and correctness of the tracking (minimization of random clusters attached to tracks). The algorithm uses as an input the ITS clusters from all six layers and

Table 1. Tracking time of HLT TPC GPU tracking processing two events in parallel, averaged over a selection of central Pb-Pb events.

Number of events	Time	Time per event
1	145 ms	145 ms
2	220 ms	110 ms

 DOI: 10.1051/epjconf/201612700014

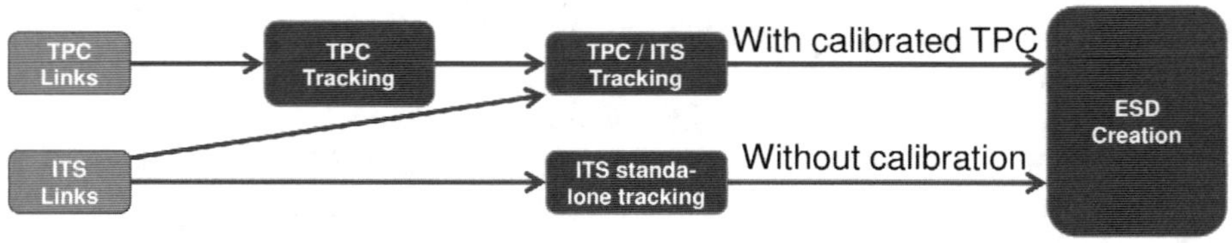

Figure 3. Scheme of ITS tracking in the HLT with two branches. The lower branch runs the ITS standalone tracer providing ITS tracks in any case. The upper branch uses TPC to ITS prolongation resulting in a higher efficiency, but it needs a calibrated TPC in order to achieve reasonable resolution.

the primary vertex provided by the HLT components running upstream. The latter is obtained as the position to which the maximum number of vectors connecting the two innermost ITS layers (silicon pixel detectors, SPD) converge. The track reconstruction starts by rebuilding these vectors (tracklets), i. e. finding pairs of SPD clusters seen under nearly the same angle from the vertex point. The procedure is the optimized version of the off-line tracklet finding described in [7].

At the following step the tracks are found by following the tracklets to ITS layers at larger radii, where the ITS Silicon Drift and Strip Detectors (SDD and SSD, respectively) are . For every tracklet a Kalman filter is initialized by the momentum estimated from the vertex and a pair of SPD clusters. It is propagated outwards starting from the vertex, considered as a measured point. The Kalman prediction/update is done first with already attached SPD clusters, then on the SDD and SSD clusters closest to extrapolation point, provided they meet a strict track-to-cluster χ^2 cut. The magnetic field is taken to be a constant solenoidal one, an approximation correct to $\approx 10^{-3}$ in the ITS volume. Multiple scattering is accounted for using the average material budget per layer. At least four clusters are required per track in the reconstruction in standard layout with all six ITS layers present. Tracks with two consecutive layers without contribution are rejected. Once the outermost layer is reached, the track *outward* kinematics is recorded (for the further matching with TPC) and inward Kalman propagation is performed to obtain the kinematics at the vertex region.

The benchmark with simulated In pp and p-Pb events, on a single core of i7-2600 CPU @ 3.40 GHz, shows a rate obove 2 kHz with a reconstruction efficiency (with respect to reconstructable Monte Carlo tracks) exceeding 90% at $p_T > 300\,\text{MeV}/c$ and a contamination by fake tracks below 3%. Minimum bias Pb-Pb events are reconstructed at a 18 Hz rate, which rises to 40 Hz if 15% of the most central collisions are skipped. The efficiency above $p_T > 300\,\text{MeV}/c$ drops to about 85%, and the fake rate is below 10%.

4.3 Luminous region estimation based on ITS tracking

The volume where the particle beams overlap and most interactions take place is called the luminous region (LR). It can only be reliably determined by the experiment itself by means of reconstructing and localizing each interaction and statistically determining the size and shape of the beam overlap volume. This measurement, performed in (quasi-) real time, is used by the LHC to optimize the beam parameters. The computation of LR requires accurate tracking close to the interaction point. TPC tracks extrapolated by about 80 cm towards the vertex have insufficient accuracy in this respect if the TPC is not fully calibrated. A more robust method is to use ITS standalone tracking which does not require large extrapolation steps or the same degree of time dependent alignment and calibration as the TPC. The ITS standalone tracker has sufficient tracking efficiency and resolution. This method is successfully implemented in the HLT and provides real time LR information to the LHC.

DOI: 10.1051/epjconf/201612700014

The vertex determination is implemented in the same component that performs the ITS standalone tracking. First the tracks are propagated to the beam-line, then a fast linear fitter is deployed to find a vertex as a point minimizing the distance of the closest approach for maximum number of tracks. The rejection of outlier tracks is achieved by means of bi-square weighting filter, as described in [8].

5 Data structures for fast data transport in the HLT

5.1 A flat data structure

As discussed in Section 2.1, one of the crucial features of the calibration procedures running in the HLT is the fact that they use the same code as used in the offline calibration. This was developed to work on the output of the offline reconstruction, i. e. on the C++ structures called Event Summary Data (ESD) on which the ALICE analyses are based. The ESDs are not suited to be used in the HLT framework since the multiple processes that may run in the HLT do not share the same address space. This implies that shipping C++ objects (possibly in the form of ROOT objects) between HLT processes can be done only through serialization and deserialization, which would introduce an unacceptable overhead in the framework for every process that accesses the ESDs. The problem that arises when tasks that take as input the standard ESDs when run offline (e. g. calibration tasks, Quality Assurance, analysis...) are run in the HLT, can be overcome by storing the output of the HLT reconstruction in flat structures, which are exchanged between the different HLT reconstruction, calibration, and QA components complying with the HLT framework requirements.

The development and implementation of the flat ESDs is based on the C++ concepts of inheritance and polymorphism. A common base class for the flat ESDs and the standard ESDs allows to run the same calibration algorithm online in the HLT and offline.

While in the ESDs the different objects that come out of the reconstruction are stored inside the ESDs as different C++ objects, in the flat ESDs they are simply stored consecutively in memory, and the bookkeeping of their position in the object is used to access them. One special case among these objects is the so-called ESD friends, which is an object meant to store information not needed for analysis, but specifically for calibration. For example, it contains the information about the clusters used to form the tracks. Since every track can have up to 159 clusters associated in the TPC (due to the same number of pad-rows in the detector), the amount of data in the friends is very large, and the size of the friends can be several times the one of the ESD tracks.

5.2 Benchmarks

We have evaluated the performance of the flat structures compared to standard ESDs with several benchmarks. These are based on the output of the HLT reconstruction when stored either in the standard ESD objects or in the flat structures. Figure 4 shows the time needed to create and manipulate (serialize, deserialize) the standard ESDs and the flat ESDs as a function of the track multiplicity of an event, based on a sample of around 500 Pb-Pb events taken in 2015. As one can see, the time for the different stages grows linearly with the track multiplicity. The time needed to create the flat ESDs including the friends is about ten times smaller than the time needed to create the standard ESDs with friends. The serialization (deserialization) of the standard ESDs including friends is two to three times (about 1.5 times) slower than their creation. The reinitialization of the flat ESDs is a step needed in order to restore the virtual table of the flat ESD object, which is needed for the common interface of ESD and flat ESD. The plot also shows the case when the standard ESDs are created and serialized without the friends, although not appropriate for calibration purposes. The time in this case is of the same order as the one needed to create the flat ESDs with the full calibration information. The behaviour of the standard ESDs without the friends compared to the flat structure is better visible in Fig. 5: also in this case the flat structures with the complete calibration information are more performant.

Figure 6 shows the time needed by the TPC drift time calibration task when running on standard ESDs and on flat ESDs. As one can see, the task performance is comparable when using as input the standard ESDs or the flat ones, being a few percent faster in case of the flat ESDs at high multiplicities. This is an additional minor advantage of the flat ESDs. The main advantage of the flat structures lies in the speed with which they are created and can then be shipped between different HLT components, more than in how fast they can be processed by the calibration task. This is anyway running asynchronously with respect to data taking, and as such it does not impose any performance limitation.

Table 2 summarizes the resources needed to create the standard ESDs and the flat ESDs (with and without friends) in the HLT in number of CPU cores. Considering the total amount of resources available in the HLT cluster, i. e. $\approx$ 2000 CPU cores, one realizes immediately that while the creation of standard ESDs as natural output of the HLT reconstruction is an expensive but well affordable process, adding the extra information needed to perform calibration tasks—i. e. what is stored in the friends—would require half of the total HLT resources, which is completely infeasible. In opposite, the flat ESDs can be created in the HLT together with the extra calibration data using a limited amount of resource, which is even smaller than those needed to generate the standard ESDs without friends.

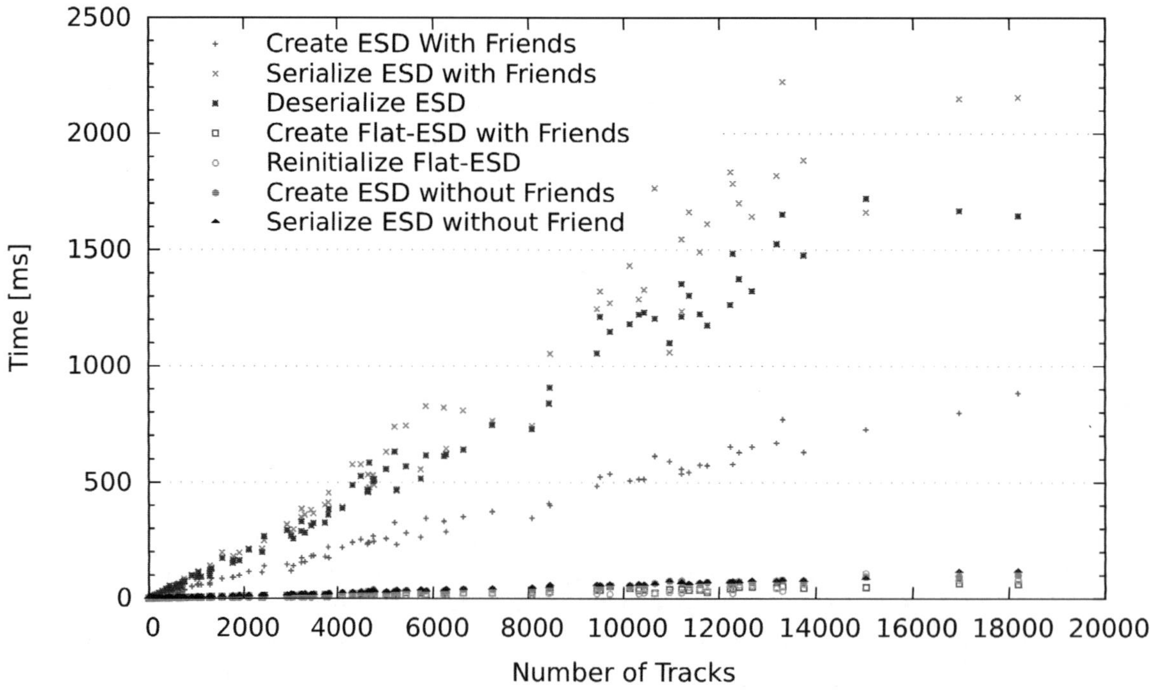

Figure 4. Time needed for different steps in the creation and manipulation of the standard and flat ESDs as a function of the track multiplicity of the event, obtained processing Pb-Pb events from 2015.

Table 2. Number of CPU cores required to create normal and flat ESDs with and without friends obtained processing Pb-Pb (from 2015) events at 300 Hz.

ESD Type	with / without friends	Number of CPU cores
ESD Converter	without friends	94.7
ESD Converter	with friends	947.0
Flat-ESD Converter	without friends	11.0
Flat-ESD Converter	with friends	72.4

6 Overview of HLT online calibration and a first online calibration test run

Figure 7 gives an overview over the data flow and all components in the HLT. The actual components used for online calibration are shown in red, and they use the new asynchronous processing feature. The actual calibration component is the only one that runs with many instances on all computing nodes. There is only one instance of the merger and the preprocessors, running on a dedicated calibration node. Then, the new TPC transformation maps based on the new calibration is distributed in the cluster via ZeroMQ. In order to make sure that the calibration component does not run on calibrated clusters, the TPC transformation component runs two instances. One uses the new transformation map from the feedback loop. The clusters from this instance are shipped to the tracker component. The other instance uses always the default transformation map and ships the clusters to the ESD, from where they go to the calibration component. The calibrated clusters in the tracks and the uncalibrated clusters for the ESD friends are identified via the cluster index which is unchanged

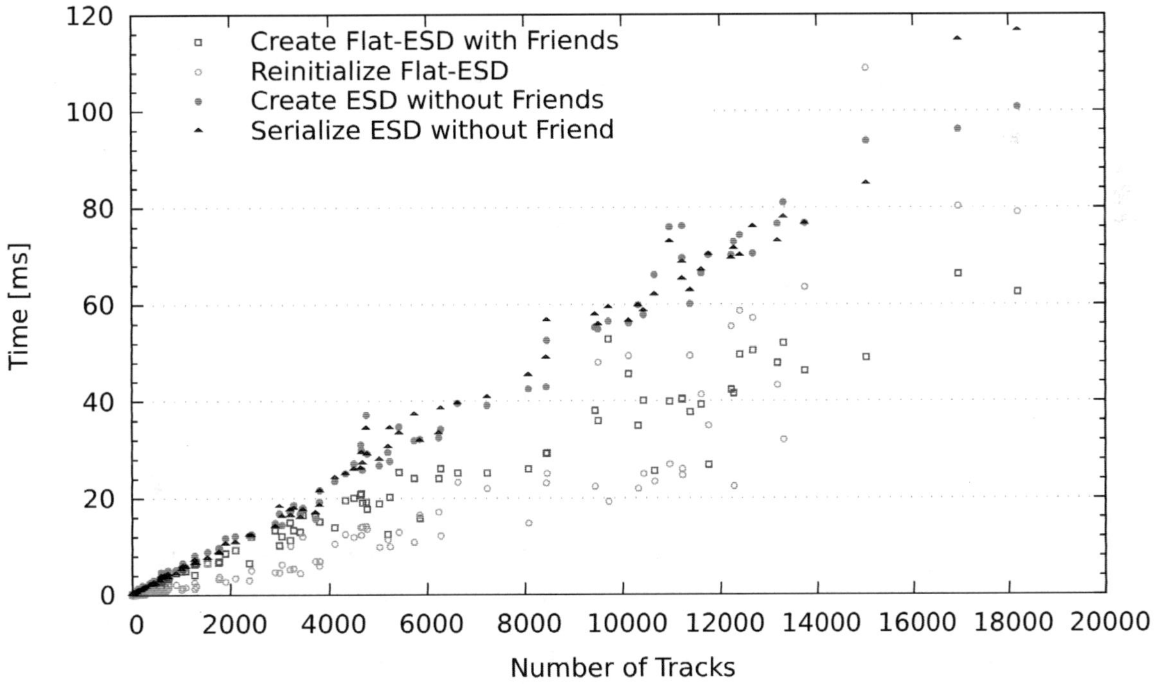

Figure 5. Time needed for different steps in the creation and manipulation of the standard and flat ESDs as a function of the track multiplicity of the event.

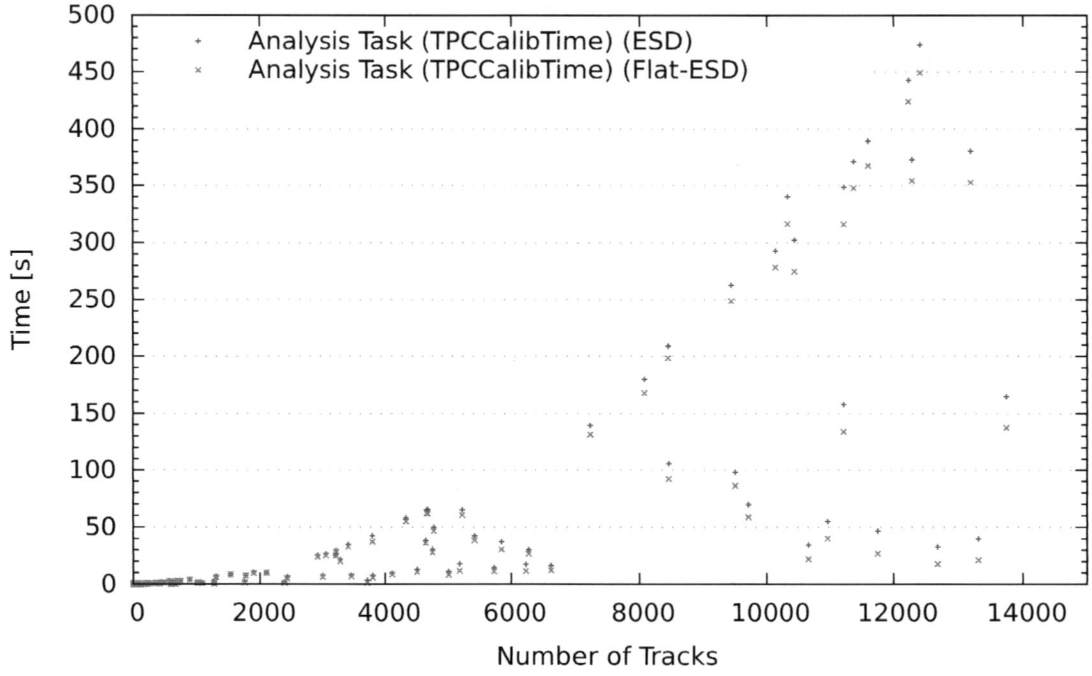

Figure 6. Time needed for different steps in the creation and manipulation of the standard (without friends) and flat ESDs as a function of the track multiplicity of the event, obtained processing Pb-Pb (from 2015) events.

by the cluster transformation. Thus, the feedback loop improves the track finding, but does not interfere with the calibration itself.

Therefore, all the requirements of online calibration in the HLT listed in Section 2.1 are met. A first test of the above-described new features with online calibration was performed in December 2015. The calibration components were running online under real conditions during Pb-Pb data taking in ALICE at LHC design luminosity. This test proved that the HLT can handle the online calibration in a high load scenario with the highest data rates.

A first analysis of the processing rate of the calibration component shows the following: the processing time of the calibration task for individual events can be quite long. In particular, it depends superlinearly on the number of tracks in the event because of the TPC–ITS matching. During this first test, it took up to 15 minutes for the largest events, which is too long in order to use the results in the feedback loop. Fortunately, the fraction of events which need more than 5 minutes for the calibration is only 2 %. In particular, it is not necessary to run the calibration for the large events at all. As it is essential to have enough tracks for the TPC–ITS matching in total, it is faster to process the same number of tracks in many smaller events compared to fewer large events. Excluding the small fraction of events that take very long, the processing of 5000 events (Step A), which is considered sufficient for the TPC drift time calibration in Pb-Pb, takes less than 5 minutes. The feedback loop (Step B) takes around 20 seconds. If the calibration is used for the following five minutes (Step C), the total time is below the limit of 15 minutes where the calibration is stable. On top of this, this first test used the plain offline calibration software. For the future, we plan to apply code optimizations to speed up the calibration task.

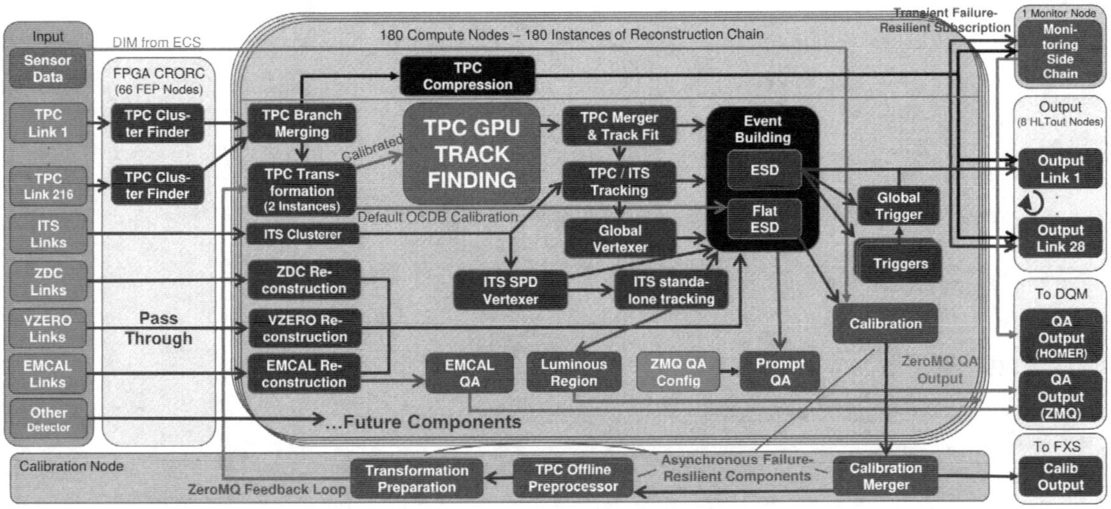

Figure 7. Overview of all processing components and data flow in the HLT.

References

[1] A. K. et al. (ALICE Collaboration), JINST **3**, S08002 (2008)

[2] C. Fabjan et al. (ALICE Collaboration), *ALICE trigger data-acquisition high-level trigger and control system: Technical Design Report*, ALICE-TDR-10; CERN-LHCC-2003-062 (CERN, Geneva, 2004), `https://cds.cern.ch/record/684651`

[3] Facility for Antiproton and Ion Research in Europe Gmb, *FAIR - Baseline Technical Report*, Vol. 1 (2006), ISBN 3-9811298-0-6, `http://www.fair-center.de/index.php?id=171&L=0`

[4] D. Rohr et al., J. Phys. Conf. Ser. **664**, 082047 (2015)

[5] S. Gorbunov et al., IEEE Trans. Nucl. Sci. **58**, 1845 (2011)

[6] D. Rohr et al., J. Phys. Conf. Ser. **396**, 012044 (2012)

[7] K. Aamodt et al. (ALICE Collaboration), Phys. Rev. Lett. **105**, 252301 (2010)

[8] B. Abelev et al. (ALICE Collaboration), Int. J. Mod. Phys. A **29**, 1430044 (2014)

A New Track Reconstruction Algorithm suitable for Parallel Processing based on Hit Triplets and Broken Lines

André **Schöning**[1,a]

[1] *Physikalisches Institut, Ruprecht-Karls Universität Heidelberg, Germany*

Abstract. Track reconstruction in high track multiplicity environments at current and future high rate particle physics experiments is a big challenge and very time consuming. The search for track seeds and the fitting of track candidates are usually the most time consuming steps in the track reconstruction. Here, a new and fast track reconstruction method based on hit triplets is proposed which exploits a three-dimensional fit model including multiple scattering and hit uncertainties from the very start, including the search for track seeds. The hit triplet based reconstruction method assumes a homogeneous magnetic field which allows to give an analytical solutions for the triplet fit result. This method is highly parallelizable, needs fewer operations than other standard track reconstruction methods and is therefore ideal for the implementation on parallel computing architectures. The proposed track reconstruction algorithm has been studied in the context of the Mu3e-experiment and a typical LHC experiment.

1 Introduction

Track reconstruction in high energy experiments at high luminosity colliders is a big challenge. The large particle rates together with the track extrapolation uncertainties, generated by multiple Coulomb scattering in the silicon tracking detector material, lead to hit confusion and define the hit combinatorial problem. High track multiplicities not only increase considerably the computational effort for solving this combinatorial problem, they also reduce the track finding efficiency and purity.

Standard track reconstruction algorithms employed in high track multiplicity environments consist typically of the following steps:

1 **Seed** search: A minimum number of hits in several tracking layers is required to form a track seed. For the reconstruction of initial track parameters at least three points are required. This could be hits in three detector layers or in only two detector layers if additional constraints on the origin of the track, e.g. collision point, are employed.

2 **Fitting** of track prototypes and seeds: By using a tracking model, which ideally should include spatial hit uncertainties as well as multiple scattering uncertainties, track parameters of seeds and prototypes[1] are determined.

[a] e-mail: schoning@physi.uni-heidelberg.de

[1] A track prototype is defined to be an incomplete track candidate which has usually more than three hits.

3 **Hit linking**: Using previously fitted track parameters, the particle trajectory of the track prototype is extrapolated to link hits from further detector layers. Track extrapolation uncertainties are usually included by defining a search window. Hit linking is terminated if hits in all tracking layers are found or no further hits can be found because of missing hits.

4 **Fit quality:** Track candidates have to fulfill some quality criteria, that is a minimum number of linked hits (layers), minimum track length, minimum track fit quality, etc.

5 **Arbitration** between track candidates: Different track candidates can share one or several hits. An arbitration algorithm can be used to find the "best" track candidate. Other track candidates sharing the same hits are discarded.

The steps 2 and 3 are iterated until all hits are linked or no more track candidates can be found. The Kalman filter [1 – 3] is an example for such an iterative algorithm which is used as standard tool for track reconstruction in many particle physics experiments.

Besides track fitting the most time consuming step is usually the track seed finding due to the hit combinatorics. For a track seed search in n_{layer} layers the processing time scales with $(N_{\mathrm{hit}})^{n_{layer}}$ where N_{hit} is the number of hits. For this reason track seed searches are often performed in regions of low occupancies (i.e. outer tracking layers in case of collider experiments). Different methods can be used to further reduce the hit combinatorial problem in the track seed search in the successive hit linking step:

- application of simple geometrical preselection cuts,
- application of more advanced algorithms like cellular automata [4],
- Hough transforms [5],
- usage of lookup tables or associative memories.

Here, a new method is proposed for solving the hit combinatorial problem for seed finding and linking of hits to tracks. It can be used to accelerate the steps 1 – 4 in comparison with other standard reconstruction algorithms.[2] The proposed method is based on hit triplets for which the track parameters can be instantly calculated by exploiting a three-dimensional fit model including multiple scattering and hit uncertainties. This fit requires the three-dimensional hit coordinates including their uncertainties and is only applicable in homogeneous magnetic fields. Both requirements are usually fulfilled in modern pixel tracking devices.

2 Method

The hit triplet based track reconstruction method works as follows. In a first step hit triplets are found by fitting three-hit combinations and by applying a χ^2-cut on the triplet fit quality. What is new compared with other similar methods is here the full exploitation of the tracking model at this very first track reconstruction step, thus reducing the number of wrong seeds and mitigating the combinatorial problem. As this method is rigorous it also works for "large triplets" with far distant hits and large track curvatures. After having found valid hit triplets, conjoined triplets (triplets which share two hits) are merged successively to full tracks, see Figure 1. The resulting trajectory is finally represented by broken lines. The track parameters are combined using a simple and fast sum which is explained in more detail in Section 3. This fast summing technique is much faster than other reconstruction methods which are based on iterations and rely on matrix inversions. Furthermore, the fast summing

[2] The arbitration step is not addressed further here.

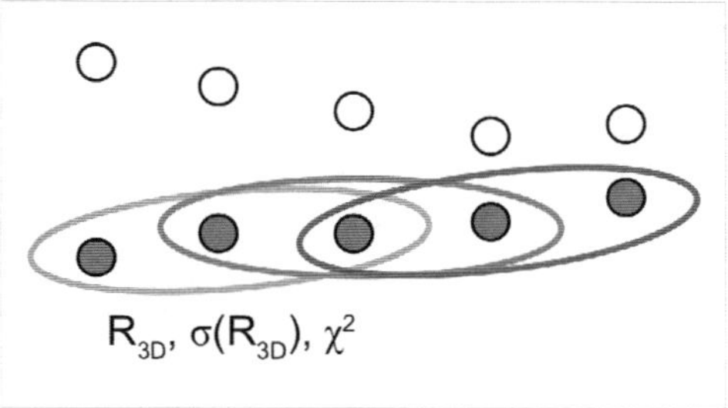

Figure 1. Sketch of merging hit triplets to full tracks. The track fit result of each triplet is given by the three-dimensional radius R_{3D}, its uncertainty $\sigma(R_{3D})$ and the fit quality χ^2.

technique is exact if multiple scattering uncertainties dominate, being the case for the majority of particles at hadron colliders if silicon tracking detectors are used.

Above described merging technique only includes local but no long-range hit correlations which are important for the reconstruction of high momentum tracks. For high momentum tracks spatial hit uncertainties dominate over multiple scattering uncertainties and a global fit is required. This can be realized by applying a General Broken Line (GBL) fit [6–8] after triplet merging. The GBL fit is an extended track fit that takes into account both, scattering effects and spatial uncertainties, and has been shown [8] to be equivalent to the Kalman filter [2]. The GBL algorithm performs a linearized fit by varying positions and kink angles at selected points around a reference trajectory, which is here chosen to be the merged triplet trajectory.

2.1 Fitting of hit triplets

The most general objective function for track fitting includes multiple scattering and spatial hit uncertainties

$$\chi^2 = \frac{\Theta_{MS}^2}{\sigma_\theta^2} + \frac{\Phi_{MS}^2}{\sigma_\phi^2} + \sum_{jk} (\mathbf{x_j} - \xi_\mathbf{j}) \mathbf{V}_\mathbf{jk}^{-1} (\mathbf{x_k} - \xi_\mathbf{k}) \,. \tag{1}$$

Here, Φ_{MS} and Θ_{MS} are the multiple scattering uncertainties in the transverse and longitudinal plane, see Figure 2. $\mathbf{x_k}$ and $\xi_\mathbf{k}$ are the fitted and measured hit coordinates and $\mathbf{V}$ is the covariance matrix.

For a particle moving in a homogeneous magnetic field an analytical solution can be calculated for a trajectory measured at three points (hit triplet). The solution for the case of negligible spatial hit uncertainties is briefly summarized in the next section (for a full description see [9]). The solution for including spatial hit uncertainties is discussed in the following Section 2.1.2.

2.1.1 Triplet fit with multiple scattering

If spatial uncertainties are negligible, i.e. $\mathbf{x_k} = \xi_\mathbf{k}$, the last term on the right-hand side of Equation (1) can be neglected. Under the assumption of energy conservation the fit has one degree of freedom.[3] The multiple scattering angles projected in the transverse and longitudinal plane can then be written as a function of the three-dimensional radius $\Phi_{MS} = \Phi_{MS}(R_{3D})$ and $\Theta_{MS} = \Theta_{MS}(R_{3D})$ which in turn is related to the momentum p of the particle via $R_{3D} = p/(qB)$ with q being the electric charge.

[3] Energy loss in silicon detectors is typically small. If required, energy loss can also be corrected for in the triplet fit procedure.

 DOI: 10.1051/epjconf/201612700015

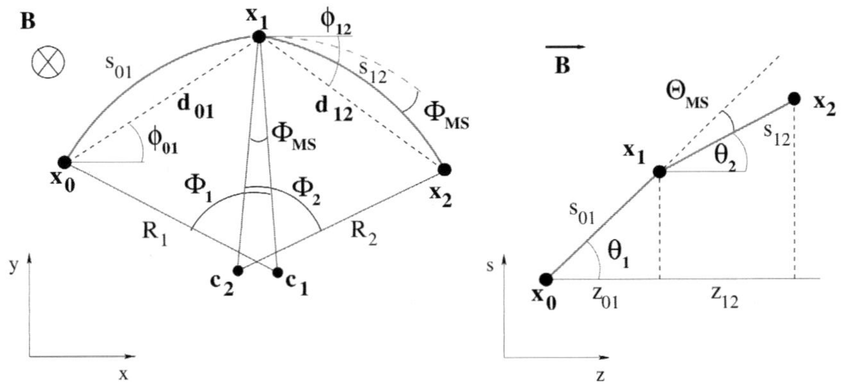

Figure 2. Particle trajectory in a homogeneous magnetic field defined by a triplet of hits $\mathbf{x_0}$, $\mathbf{x_1}$ and $\mathbf{x_2}$. The left view shows the projection to the plane transverse to the magnetic field, whereas the right view is a projection to the field axis-arclength (s) plane. Φ_{MS} and Θ_{MS} denote the multiple scattering angles in the transverse and longitudinal plane.

$\Phi_{MS}(R_{3D})$ and $\Theta_{MS}(R_{3D})$ are transcendent functions which can not be algebraically solved. However, the minimization condition

$$\frac{d}{dR_{3D}}\chi^2(R_{3D}) = 0$$

can be solved by linearization around the circle solution in the transverse plane which is described in detail in [9]. The result[4] for the three-dimensional radius

$$R_{3D} = -\frac{\eta\,\tilde{\Phi}_C\,\sin^2\vartheta_C + \beta\,\tilde{\Theta}_C}{\eta^2\,\sin^2\vartheta_C + \beta^2} \tag{2}$$

depends only on parameters derived from the relative position of the hit coordinates, $\eta, \tilde{\Phi}_C, \beta, \tilde{\Theta}_C = f(\mathbf{x_k})$, and the polar angle ϑ_C which can be approximated from the circle solution. The exact definitions of these geometric parameters can be found in [9]. The corresponding uncertainty is given by

$$\sigma(R_{3D}) = \sigma_{MS}\sqrt{\frac{1}{\eta^2\,\sin^2\vartheta_C + \beta^2}}$$

and contains the expected multiple scattering uncertainty, which can be calculated using the Highland approximation [10]

$$\sigma_{MS} \approx \frac{13.6\ \mathrm{MeV/c}}{\beta p}\sqrt{X/X_0} \tag{3}$$

after fitting, using the effective material thickness X at the middle hit position where the scattering takes place. Finally the fit quality is given by

$$\chi^2_{\min} = \frac{1}{\sigma^2_{MS}}\frac{(\beta\,\tilde{\Phi}_C - \eta\,\tilde{\Theta}_C)^2}{\eta^2 + \beta^2/\sin^2\vartheta_C}\,.$$

[4] In Reference [9] a second formula for R_{3D} is given which includes a bias correction. For sake of simplicity we use here the result from Equation 23 in [9].

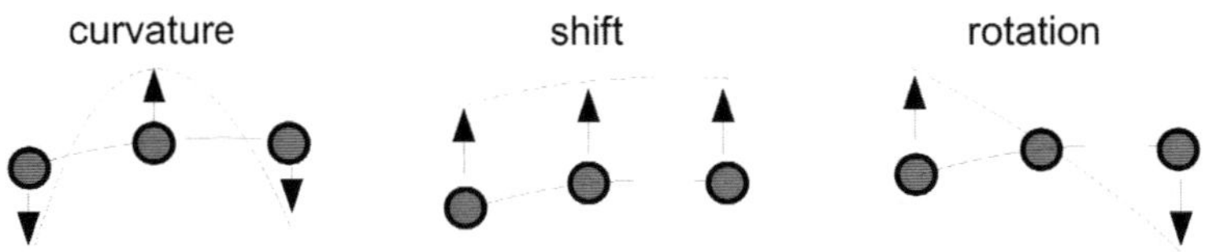

Figure 3. Representation of hit uncertainty modes of a hit triplet. From left to right: curvature uncertainty, shift uncertainty and rotation uncertainty.

2.1.2 Triplet fit with hit uncertainties

In case of non-negligible spatial hit uncertainties the above description yields too large χ^2-values and will fail. For hit triplets this bias can be fixed elegantly by absorbing the hit uncertainties in the scattering angles. By introducing effective scattering angles

$$\Delta\Theta = \Theta_{\mathrm{MS}} + \Theta_{\mathrm{hit}} \quad \text{and} \quad \Delta\Phi = \Phi_{\mathrm{MS}} + \Phi_{\mathrm{hit}} ,$$

a new objective function can be defined

$$\chi^2 = \frac{\Delta\Theta^2}{\sigma^2_{\theta_{\mathrm{eff}}}} + \frac{\Delta\Phi^2}{\sigma^2_{\phi_{\mathrm{eff}}}},$$

where effective hit uncertainties have been introduced

$$\sigma^2_{\theta\mathrm{eff}} = \sigma^2_{\mathrm{MS}} + \sigma^2_{\theta_{\mathrm{hit}}} \quad \text{and} \quad \sigma^2_{\Phi\mathrm{eff}} = \frac{\sigma^2_{\mathrm{MS}}}{\sin^2\vartheta_C} + \sigma^2_{\Phi_{\mathrm{hit}}}.$$

The hit-related uncertainties $\sigma_{\theta_{\mathrm{hit}}}$ and $\sigma_{\Phi_{\mathrm{hit}}}$ are obtained by propagating the uncertainties from the hit positions, see Appendix A. Note that this simplified procedure does not fit the hit positions, i.e. $\mathbf{x_k} = \xi_\mathbf{k}$, but it yields the correct χ^2-distribution, i.e. the probability distribution is flat. This is essential to validate the hit triplets for the track reconstruction.

Finally, the three-dimensional radius is obtained by calculating:

$$R^{\mathrm{min}}_{\mathrm{3D}} = -\frac{\eta\,\tilde{\Phi}_{\mathrm{C}}\sin{\vartheta_C}^2/\sigma_\phi(R^{\mathrm{min}}_{\mathrm{3D}})^2 + \beta\,\tilde{\Theta}_{\mathrm{C}}/\sigma_\theta(R^{\mathrm{min}}_{\mathrm{3D}})^2}{\eta^2\sin{\vartheta_C}^2/\sigma_\phi(R^{\mathrm{min}}_{\mathrm{3D}})^2 + \beta^2/\sigma_\theta(R^{\mathrm{min}}_{\mathrm{3D}})^2}.$$

By using the Highland approximation (Equation (3)) above equation can be transformed into a cubic equation in $R^{\mathrm{min}}_{\mathrm{3D}}$ which is algebraically solvable, see Appendix C.

The inclusion of the hit uncertainties generates nine extra degrees of freedom. Relevant for track fitting and reconstruction are only six[5] , which can be represented as "triplet bending", "triplet shifting" and "triplet rotation" in the transverse as well as in the longitudinal plane, see Figure 3. The triplet bending is included in above fitting procedure, whereas shifts and rotations of the hit triplet are not. For track extrapolations, however, they might be relevant and have to be included. The rotational uncertainty of a hit triplet is given in Appendix B.

[5] The other three degrees of freedom correspond to two oscillation modes and a shift in longitudinal direction, which are completely negligible because of their small size.

3 Triplet linking and merging

For the linking of hit triplets (or between a validated triplet and a track prototype) two strategies can be employed, the intersection technique and the extrapolation technique. In the intersection technique hits shared between validated triplets are used to identify connected tracks, see Figure 1. This can be done quickly in hardware by using fast lookup techniques or in software by using binary search trees, for example.

The extrapolation technique extrapolates already reconstructed track prototypes or triplets into other detector layers to add new single hits or hit triplets. This method is very efficient if the search window is small. That is typically the case for long track prototypes for which the track parameters have been precisely fitted in the previous step. Both techniques can also be combined to speed up linking.

The consistency of a new track hypothesis can be tested by calculating a combined χ^2_{comb} value by summing over all contributing triplets, where the number of hit triplets given is by $n_{\text{hit}} - 2$. There is some freedom in how to perform the weighting in the sum. For dominating multiple scattering uncertainties and if the $R_{\text{3D,i}}$ were calculated using Equation (2) the optimal sum reads:

$$\chi^2_{\text{comb}} = \sum_i^{n_{\text{hit}}-2} \chi_i^2 + \frac{(R_{\text{3D,i}} - \overline{R_{\text{3D}}})^2}{\sigma_i(\overline{R_{3D}})^2} \tag{4}$$

with

$$\overline{R_{3D}} = \sum_i^{n_{\text{hit}}-2} \frac{R^2_{3D,i}}{\sigma(R_{3D,i})^2} \Big/ \sum_i^{n_{\text{hit}}-2} \frac{R_{3D,i}}{\sigma(R_{3D,i})^2}. \tag{5}$$

This definition of the averaged three-dimensional radius is free of bias. Equations 4 and 5 are exact and no refitting of the track parameters after the linking step is required.

For dominating hit uncertainties, the relative momentum resolution is no longer constant and it is advantageous to perform the χ^2 sum and the averaging in three-dimensional curvature, $h_{3D} = 1/R_{3D}$, yielding:

$$\chi^2_{\text{comb}} = \sum_i^{n_{\text{hit}}-2} \chi_i^2 + \frac{(h_{3D,i} - \overline{h_{3D}})^2}{\sigma_i(\overline{h_{3D}})^2} \tag{6}$$

with

$$\overline{h_{3D}} = \sum_i^{n_{\text{hit}}-2} \frac{h_{3D,i}}{\sigma(h_{3D,i})^2} \Big/ \sum_i^{n_{\text{hit}}-2} \frac{1}{\sigma(h_{3D,i})^2}. \tag{7}$$

Equations 6 and 7 are not exact since long range correlations are ignored and triplet parameter uncertainties are overestimated in the procedure described in Section 2.1.2. The so obtained combined χ^2 value is only a lower (optimistic) estimate. However, this combination procedure is very fast, highly parallelizable and useful for track fitting as it allows to quickly reject wrong triplet combinations without refitting.

The full precision is finally obtained by fitting, in addition to the kink angles, the hit positions $\mathbf{x}$, see next section.

4 General Broken Line fit

Long-range correlations among different hit triplets can be included by the GBL fit [6–8]. It performs a linearization around a reference trajectory which is here chosen to be the broken line solution

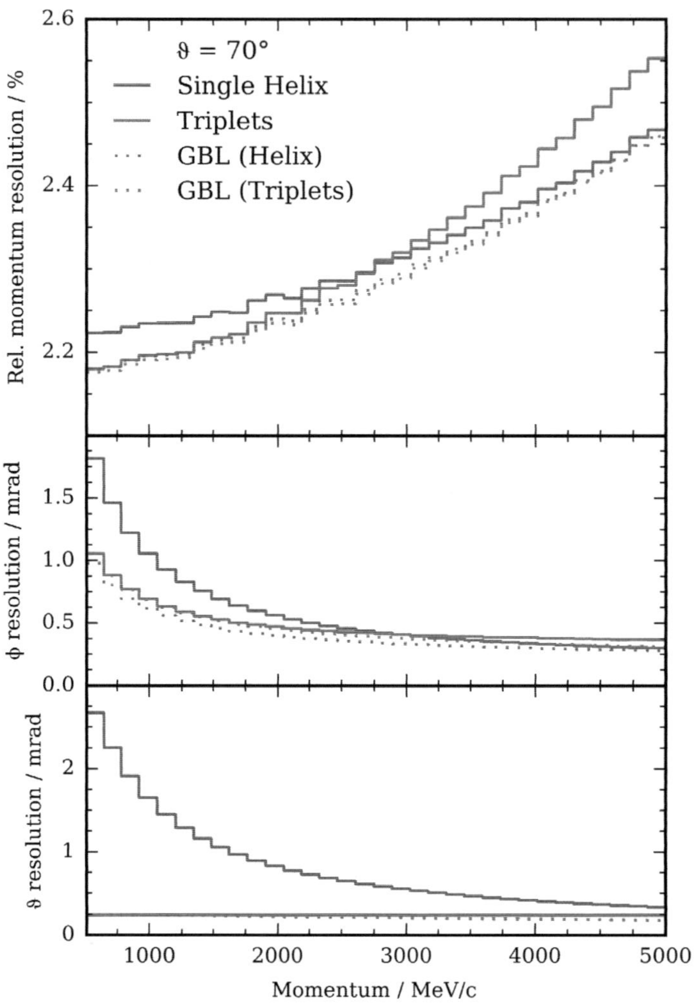

Figure 4. Track parameter resolution for an equidistant layer geometry of a high energy LHC-type detector as described in [9]. The top panel shows the relative momentum resolution and the bottom panels show the resolution of the azimuthal angle ϕ and the polar angle ϑ for tracks with a polar angle of $\vartheta = 70°$. Plot taken from [9].

discussed in the previous section. A GBL fit using the triplets fit as reference has been recently implemented in [11, 12].

The performance of this fit and the improvement with respect to the triplets fit solution and a solution obtained from a single helix fit has been studied in [9] for a typical high energy LHC detector with equidistant layer spacing. The track parameter resolutions obtained by the different track fits for particles at $\vartheta = 70°$ are shown in Figure 4 for the momentum, azimuthal angle and polar angle. In the momentum range up to about 2 GeV/c the triplets fit provides a resolution which is equally good as the GBL fits. In this region the triplet fit is considered to be adequate. Note that more than 99% of all charged particles produced at the LHC have a transverse momentum below 1 GeV.

In the momentum range of above 2 GeV/c the track parameter resolutions is significantly improved by using the GBL fit. For even higher momenta $p \geq 5$ GeV/c a single helix fit can be applied which requires far less computational effort than the GBL fit. The combination of the triplets fit for low momenta and a single helix fit for high momenta provide a very efficient scheme for fast track reconstruction. Only in the intermediate regime between about 2 to 5 GeV/c significant performance improvements are possible by using the GBL fit.

The triplet based track reconstruction method has been successfully implemented for the Mu3e-experiment [13, 14] and tested in simulations. The Mu3e-experiment will search for the lepton flavor violating decay of stopped muons $\mu^+ \rightarrow e^+e^+e^-$. Because of the very low momentum of the decay electrons tracking is performed in kinematic region which is dominated by multiple scattering and spatial hit uncertainties can be ignored. For the Mu3e-experiment the above described track reconstruction based on hit triplets was successfully implemented on GPUs.

Triplet based tracking was also implemented for an upgraded ATLAS detector designed for High Luminosity LHC. It was successfully tested for different detector geometries and different layer spacings. A publication of the results is foreseen in the near future.

5 Summary and outlook

A new track reconstruction method based on hit triplets has been proposed. It is based on a multiple scattering based track model [9] which has been extended here to include also spatial hit uncertainties. The described triplet fit has an algebraic solution and can be used to quickly validate hit combinations. It is therefore ideally suited for track reconstruction in general, and track seed searches in particular. The proposed method gives a simple prescription for the combination of hit triplets. The fit quality and the track momentum can be represented by sums, which can be quickly calculated. This method can be easily combined with a GBL fit to improve the track parameters for non-vanishing spatial hit uncertainties. For high momentum tracks (negligible multiple scattering) a single helix fit can be used alternatively, which is much faster than the GBL fit.

The whole track reconstruction procedure is non-iterative. All hit triplets can be reconstructed in parallel, what makes it ideal for the implementation on parallel computing architectures.

The used tracking model relies on the assumption that the magnetic field within a triplet is homogeneous and that multiple scattering occurs only at or near the detection layers. The effect of non-homogeneous magnetic fields, particles with missing hits in layers due to detector efficiencies, and particle scattering in material outside the detection layers will be subject of further studies.

Acknowledgments

The author thanks Moritz Kiehn for helping to implement the GBL fit and Frank Meier-Aeschbacher for fruitful discussions and reading of the manuscript.

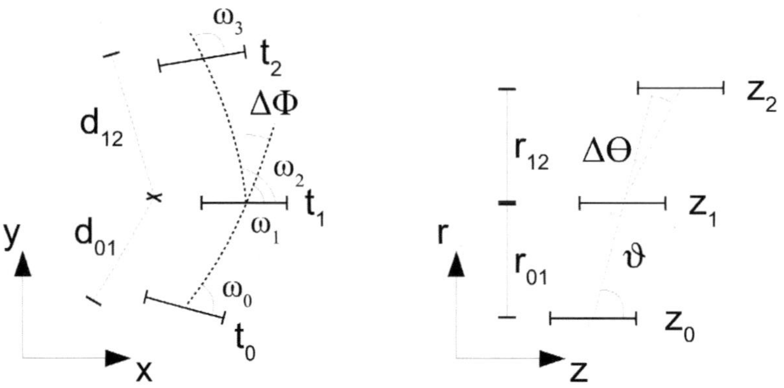

Figure 5. Hit triplet with spatial hit uncertainties and definition of the kink angle uncertainties in the transverse (left) and longitudinal (right) plane.

A Propagation of triplet hit uncertainties

A simple method to include hit uncertainties in the triplet fit is to interpret the hit residuals as kink angles: $\Delta\Phi$ in the transverse and $\Delta\Theta$ in the longitudinal plane, see Figure 5. The tilt angles with respect to the particle trajectory are given by the angles ω_i. The uncertainties in radial direction σ_{r_k}, in azimuthal direction σ_{t_k} and in axial direction σ_{z_k} can be propagated to define effective kink angle uncertainties $\sigma_{\phi_{\text{hit}}}$ and $\sigma_{\theta_{\text{hit}}}$:

$$\begin{aligned}\sigma^2_{\phi_{\text{hit}}} &= \frac{\sin^2\omega_0}{\Delta d^2_{01}}\sigma^2_{t_0} + \frac{\sin^2\omega_3}{\Delta d^2_{12}}\sigma^2_{t_2} + \left(\frac{\sin\omega_1}{\Delta d_{01}} + \frac{\sin\omega_2}{\Delta d_{12}}\right)^2\sigma^2_{t_1} \\ &+ \frac{\cos^2\omega_0}{\Delta d^2_{01}}\sigma^2_{r_0} + \frac{\cos^2\omega_3}{\Delta d^2_{12}}\sigma^2_{r_2} + \left(\frac{\cos\omega_1}{\Delta d_{01}} + \frac{\cos\omega_2}{\Delta d_{12}}\right)^2\sigma^2_{r_1}\end{aligned}$$

$$\begin{aligned}\sigma^2_{\theta_{\text{hit}}} &= \sin^4\vartheta\left(\frac{\sigma^2_{z_0}}{\Delta r^2_{01}} + \frac{\sigma^2_{z_2}}{\Delta r^2_{12}} + \sigma^2_{z_1}\left(\frac{1}{\Delta r_{01}} + \frac{1}{\Delta r_{12}}\right)^2\right) \\ &+ \sin^2\vartheta\,\cos^2\vartheta\left(\frac{\sigma^2_{r_0}}{\Delta r^2_{01}} + \frac{\sigma^2_{r_2}}{\Delta r^2_{12}} + \sigma^2_{r_1}\left(\frac{1}{\Delta r_{01}} + \frac{1}{\Delta r_{12}}\right)^2\right).\end{aligned}$$

B Triplet rotation uncertainty

For non-vanishing spatial hit uncertainties rotational degrees of freedom in the transverse and in the longitudinal plane are generated. These rotational uncertainties are important for track extrapolations. For a barrel-type detector and negligible track curvatures the rotational uncertainties are given by:

$$\begin{aligned}\delta\theta^2_{rot} &= \sin^4\vartheta\,\frac{\sigma^2_{z_0}\sigma^2_{z_1} + \sigma^2_{z_1}\sigma^2_{z_2} + \sigma^2_{z_0}\sigma^2_{z_2}}{\sigma^2_{z_0}r^2_{12} + \sigma^2_{z_2}r^2_{01} + \sigma^2_{z_1}r^2_{02}} \\ \delta\phi^2_{rot} &= \frac{\sigma^2_{t_0}\sigma^2_{t_1} + \sigma^2_{t_1}\sigma^2_{t_2} + \sigma^2_{t_0}\sigma^2_{t_2}}{\sigma^2_{t_0}d^2_{12} + \sigma^2_{t_2}d^2_{01} + \sigma^2_{t_1}d^2_{02}}.\end{aligned}$$

Here the same notation as in Appendix A is used. It is straight-forward to calculate the rotational uncertainties for other detector orientations.

C Extended triplet fit with hit uncertainties

The three-dimensional radius in an extended triplet fit including hit uncertainties is given by:

$$R_{3D}^{\min} = -\frac{\eta\,\tilde{\Phi}_C \sin\vartheta^2/\sigma_\phi(R_{3D}^{\min})^2 + \beta\,\tilde{\Theta}_C/\sigma_\theta(R_{3D}^{\min})^2}{\eta^2 \sin\vartheta^2/\sigma_\phi(R_{3D}^{\min})^2 + \beta^2/\sigma_\theta(R_{3D}^{\min})^2}.$$

The polar angle of the track ϑ can be taken from the circle solution, which also can serve as a first estimate to calculate the inclination angles ω_i in the transverse plane, see Figure 5. Note that in general $\sigma_\phi \neq \sigma_\theta$.

A difficulty arises from the fact that the multiple scattering uncertainty depends on the particle momentum, which is not known before track fitting. A solution is to include this momentum dependence in the fit by using the Highland parametrization (Equation (3)), leading to the following cubic equation:

$$\begin{aligned} 0 &= {R_{3D}^{\min}}^3\,(\eta^2\sigma^2_{\theta_{\rm hit}} + \beta^2\sigma^2_{\phi_{\rm hit}}) \\ &+ {R_{3D}^{\min}}^2\,(\eta\,\tilde{\Phi}_C\,\sigma^2_{\theta_{\rm hit}} + \beta\,\tilde{\Theta}_C\,\sigma^2_{\phi_{\rm hit}}) \\ &+ R_{3D}^{\min}\,b^2\,(\eta^2 + \beta^2/\sin\vartheta^2) \\ &+ b^2\,(\eta\,\tilde{\Phi}_C + \beta\,\tilde{\Theta}_C/\sin\vartheta^2) \\ &:= a_3{R_{3D}^{\min}}^3 + a_2{R_{3D}^{\min}}^2 + a_1 R_{3D}^{\min} + a_0. \end{aligned}$$

Here b is an effective scattering parameter, which is a measure for the material thickness at the scattering layer, see [9] for definition. Using Cardano's formula it is straight forward to find the solution for $R_{3D}^{\min}$.

References

[1] R.E. Kalman, J. Basic Eng. **82**, 35 (1960)
[2] R. Frühwirth, Nucl. Inst. & Meth. A **262**, 444 (1987)
[3] P. Billoir, S. Qian, Nucl. Inst. & Meth. A **294**, 219 (1990)
[4] I. Abt, D. Emelyanov, I. Kisel, S. Masciocchi, Nucl. Inst. & Meth. A **489**, 389 (2002)
[5] P. Hough, U.S. Patent 3,069,654 (1962)
[6] V. Blobel, Nucl. Inst. & Meth. A **566**, 14 (2006)
[7] V. Blobel, C. Kleinwort, F. Meier, Comp. Phys. Comm. **182**, 1760 (2011)
[8] C. Kleinwort, Nucl. Inst. & Meth. A **673**, 107 (2012)
[9] N. Berger, M. Kiehn, A. Kozlinskyi, A. Schöning, submitted to Nucl. Inst. & Meth. **A** (2016), `https://arxiv.org/abs/1606.04990`
[10] V.L. Highland, Nucl. Inst. & Meth. A **129**, 497 (1975)
[11] M. Kiehn, Ph.D. thesis, Physics Institute, Heidelberg University (Februar 2016), `pdf`
[12] M. Kiehn, WATSON A Tracking Library, internal software, University Heidelberg
[13] A. Blondel et al. (2013), `http://arxiv.org/abs/1301.6113`
[14] N. Berger et al., Nucl. Inst. & Meth. A **732**, 61 (2013)

The Neuro-Z-Vertex Trigger of the Belle II Experiment

Sebastian **Skambraks**[1,a], Sara **Neuhaus**[1], and Christian **Kiesling**[2]

[1] *Technische Universität München, Germany*
[2] *Max-Planck-Institut für Physik, München, Germany*

Abstract. A neural network z vertex trigger is planned for the upcoming Belle II detector at the SuperKEKB collider. This neural algorithm is based on a single track 3D parameter estimation using only hit and drift time information from the central drift chamber. Due to the high luminosity ($\mathcal{L} = 8 \times 10^{35}\,\mathrm{cm}^{-2}\,\mathrm{s}^{-1}$) Belle II will have to face high levels of beam induced background, making a z vertex reconstruction at the first level trigger mandatory. Using the neural z vertex algorithm, the requirements of the standard track trigger can be strongly relaxed. By this, the trigger efficiencies, especially for low multiplicity events, e.g. τ pair production, can be significantly increased.

This contribution presents the foreseen neural network trigger setup and the preceding 2D track finder. Special focus is put on the proposal and evaluation of a possible 3D upgrade of the 2D track finder. Additionally, details are given on a dedicated setup for the upcoming cosmic ray test.

1 Introduction

The Belle II experiment [1] is an upgrade of the successful Belle [2] flavor physics experiment. It is located in Tsukuba, Japan at the upgraded asymmetric $e^+ e^-$ collider SuperKEKB. With beam energies of 4 GeV for e^+ and 7 GeV for e^- the collider runs at the center of mass energy of the $\Upsilon(4S)$ resonance. The design luminosity of SuperKEKB is $\mathcal{L} = 8 \times 10^{35}\,\mathrm{cm}^{-2}\,\mathrm{s}^{-1}$ which is 40 times larger than the world record luminosity reached by its predecessor KEKB. Typical Υ(4S) events will have an average track transverse momentum (p_T) of 500 MeV and an average multiplicity of 11 tracks per event.

Belle II consists of a high precision pixel detector (PXD) as innermost detector, followed by the silicon vertex detector (SVD). Due to the latency and data bandwidth, these high precision measurements are only available for offline reconstruction. The third detector shell is the central drift chamber (CDC), a wire chamber with 56 layers of sense wires. The CDC is the main tracking device of Belle II. With its high speed readout it is perfectly suited for the first trigger level and it provides the input to the z vertex trigger. The outer detector shells comprise calorimeters (ECL), Cherenkov counters (A-RICH), time-of-propagation counters (TOP) and a kaon and muon detecor (KLM). More details on the Belle II experiment can be found in [1].

[a] e-mail: sebastian.skambraks@tum.de

 DOI: 10.1051/epjconf/201612700016

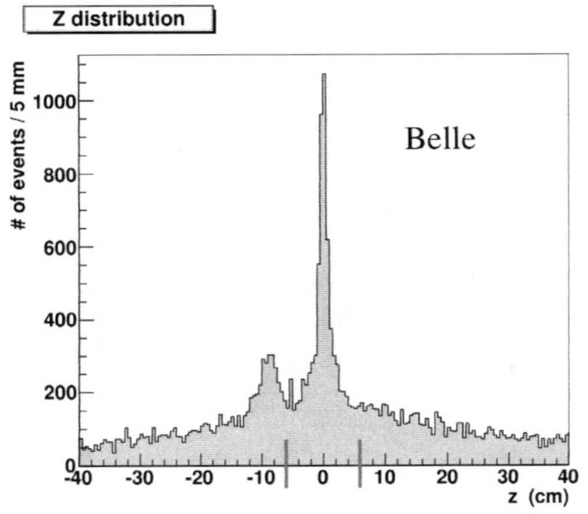

Figure 1. z distribution of random triggered events in Belle [1]. A z vertex trigger with an accuracy of $\sigma_z = 2\,\text{cm}$ would enable a cut at the red marks ($z = \pm 6\,\text{cm}$) to increase the signal to background ratio.

1.1 Background in Belle II

With the luminosity upgrade, the level of beam background will increase in Belle II. Due to the high beam collimation (nano-beam-scheme [1]), the Touschek effect [3] will become a dangerous new source for background tracks. Other dominant sources of beam background are radiative Bhabha scattering, beam–gas interactions and synchrotron radiation. Typically, background interacts with the beam pipe which results in tracks with displaced z vertices.

Here, the coordinate system to describe the vertex position has its origin at the interaction region, where the e^+ and e^- bunches are brought to collision. The z-axis is oriented along the beam line and the magnetic field points in the z direction. Charged tracks are described as a helix $(p_T, \varphi, \vartheta, z, d)$, where the track-vertex is located at z with a displacement d to the z-axis (position of closest approach). p_T is the transverse momentum, φ the azimuthal angle, and ϑ the polar angle of the track momentum at the vertex position. Displacements along the z-axis can be large compared to the radius of the beampipe ($r \approx 1\,\text{cm}$). Therefore, the displacement d are neglected at the z vertex trigger level.

The background situation in the predecessor Belle, which had no z vertex trigger, is shown in Figure 1. Only the peak around $z = 0\,\text{cm}$ corresponds to the interesting physics signal, and a z vertex trigger would allow to suppress the background tails. With the anticipated accuracy of $\sigma_z = 2\,\text{cm}$, a cut at $\pm 6\,\text{cm}$ (red marks in Figure 1) is possible while maintaining a high efficiency for the physics signal. In Belle II, a z vertex trigger at the first trigger level becomes necessary, to provide a high physics signal rate in spite of the increased background level.

1.2 τ events

Events with a low track multiplicity are most in danger to be mistaken as background. Here a z vertex trigger can significantly improve the efficiency by relaxing the other trigger conditions and otherwise unseen events can be recovered. τ events have a large cross section in Belle II and are a typical candidate for low multiplicity events. A simulation study shows the possible improvement for τ decays

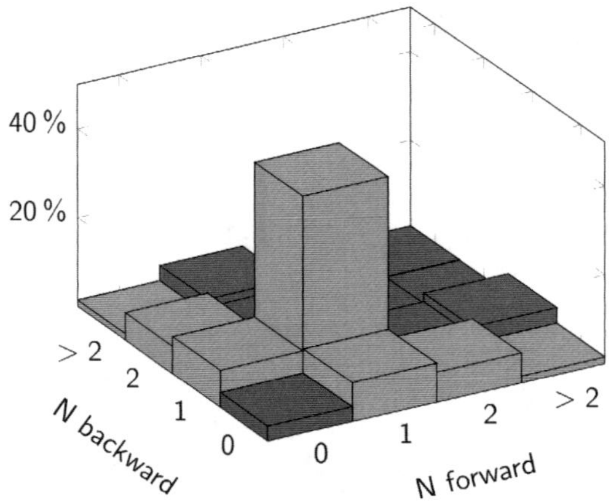

Figure 2. Simulation study of track numbers in hemispheres with τ events as an example for expected low multiplicity events. With a z vertex trigger the otherwise lost green columns can be recovered. This results in a potential efficiency gain by a factor of ≈ 3.9.

in Belle II for a trigger setup with a perfect z vertex trigger compared to a trigger setup without a z trigger. Without the z trigger at least 3 tracks are required with at least one track in each hemisphere. With a z vertex trigger, only two tracks are required with the additional condition $|z| < 6$ cm. Figure 2 shows the track counting in the forward and backward hemispheres of the Belle II detector in a pure simulation study. Without the z vertex trigger only the events in the blue columns can be triggered. With the z vertex trigger all the green columns can additionally be recovered. This corresponds to a factor of ≈ 3.9 for the total potential efficiency increase.

1.3 CDC first level trigger system

To fulfill the main requirements [1] of the first level trigger (30 kHz output rate, 5 µs latency, 200 ns event separation) only detector components with a fast readout time can be used. This leaves the CDC as the only tracking and vertexing device at the first trigger level.

A schematic of one half of the CDC is shown in Figure 3. The CDC has 14336 sense wires in 56 layers which are combined into nine superlayers (SL) of wires with the same orientation. There are five axial SLs, oriented parallel to the beam line and the magnetic field (z-axis), and four inclined stereo SLs. The different stereo angles with respect to the z-axis make a 3D reconstruction possible.

The following methods will be integrated in the pipeline of the CDC sub-trigger system. After digitization of the sense wire information of the central drift chamber, the first component is the track segment finder (TSF). It compresses the wire hit information and removes noise by combining the wire layers within the SLs to a total of 2336 track segments (TS). The TSF selects TSs as predefined patterns of wires within a SL (see Figure 3): if at least one wire in four out of the five layers within the TS pattern have a hit, then the TS is hit. The TSF outputs the id of the TS, a priority wire within the TS, and the drift time with respect to the priority wire. Additionally, it provides left/right information based on the wire hit pattern within the TS.

The TSF is followed by the track finder which is based on a Hough transformation [4]. The currently foreseen 2D track finder uses only inputs from the axial SLs. An extended 3D track finder is described in Section 3. The 2D finder provides 2D track parameter estimates (p_T, φ) for all tracks found.

DOI: 10.1051/epjconf/201612700016

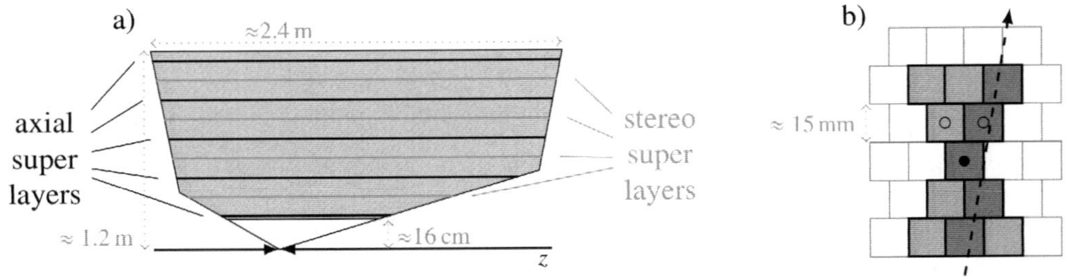

Figure 3. a) The superlayer (SL) structure of the Belle II central drift chamber (CDC). SLs with axial orientation (parallel to the z-axis) alternate with inclined stereo SLs. b) The shape of one track segment (TS), combining 6 wire layers within one SL, is shown as a colored region. The wire cells hit by the track are colored red. A track leaving a hit in at least four out of the five wire layers within the track segment shape is recognized as a TS hit. The TS hit consists of the TS id, the drift time at the priority wire and a left/right information with respect to the priority wire. The central wire has the first priority; if it is missing, one of the two wires above it is used.

All TSs and the 2D track estimates are then used as input to the 3D track reconstruction (ϑ, z). This includes the neural network, which is stable even in the presence of noise and nonlinearities occuring under realistic conditions, for example a nonlinear drift time to distance relation.

Finally, the results of the CDC sub-trigger system are fed into the global decision logic of the first level trigger.

2 Neural network for track parameter estimation

The general ideas for the neural network track parameter estimation in the Belle II experiment were developed in [5–8]. More details on the setup of the single networks, the input transformation and the specialization of networks to phase space sectors can be found there.

2.1 Network setup and training

Neural networks are powerful machine learning tools, that can be used for general function approximation [9]. In the feed-forward setup used here [8], the network has the structure of a directed, acyclic graph, where each node computes the weighted sum over its inputs and evaluates this with an activation function (hyperbolic tangent). This structure is well suited for an implementation in parallel hardware and has a deterministic runtime. Here, the neural network is used for single track z vertex estimation (see Figure 4). A preprocessing step is required to select the associated hits for each track and transform the raw hit information to the input values for the network.

The neural network receives one TS per SL per track as input. One TS is represented by two values for the position and one value for the drift time. To match the possible values of the hyperbolic tangent, all input and output values are scaled to the interval $[-1, 1]$. The floating point output value is then directly interpreted as the scaled z vertex value of the single track. The training is carried out with the iRPROP^{-} [10] backpropagation algorithm on a simulated dataset. As soon as the detector data is available, a retraining with real tracks is planned, where the precise offline reconstruction values are used as training targets.

The required preprocessing for the neural network comprises the association of TS hits to tracks and the representation of the hits in appropriate coordinates. The TS selection is currently based

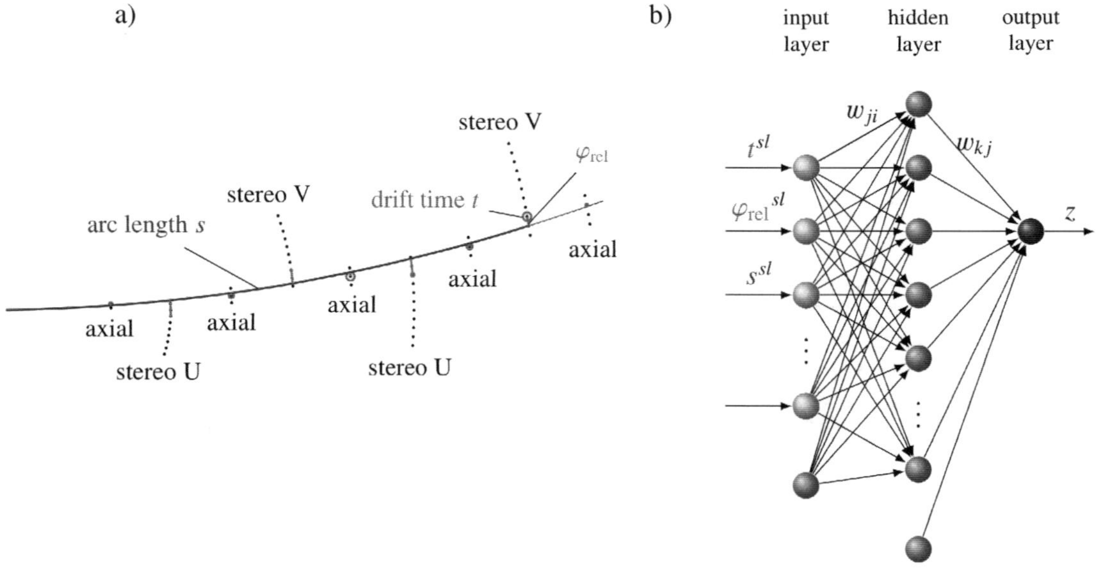

Figure 4. a) Input selection for the neural networks. Hits in small intervals around the found 2D tracks are selected and transformed to the input for the neural network. b) Structure of a neural network receiving 3 input values per superlayer (SL).

solely on the 2D track parameter estimates provided by the Hough finder. Small intervals around the 2D track estimates are defined for each SL in a pre-training run (see Figure 4). Due to the different stereo angles, the intervals of the stereo SLs are asymmetric and larger than the intervals of the axial SLs. If more than one hit is contained within the interval, the hit with the shortest drift time is used. After the hit selection, the hit positions are represented in a unique way by s and φ_{rel}, using the 2D information: the parameter s is the arc length along the 2D track to the hit, and φ_{rel} is the distance in φ of the 2D track to the hit (see Figure 4). The third component is the drift time t, where the left/right information of the TS is used as the sign. More details on the input representation can be found in [8].

To address asymmetries in ϑ, p_T and charge, the possibility for sectorization of the phase space is introduced. Sectorization in p_T and charge is possible with the present 2D finder alone, whereas ϑ sectorization is only possible if an additional ϑ estimate is introduced. One option for ϑ estimation is the 3D track finder described in Section 3. In [7] it was shown that many "expert" neural networks, specialized to small sectors of the phase space, help to improve the z estimation accuracy. Because sectorization costs memory resources and introduces a sensitive dependency on the sector selection accuracy, a small total number of sectors is desirable.

Figure 5 shows the z estimation capabilities of a basic setup with only two neural networks for different charges with a 90% RMS as resolution measure. For the realistic events with noise, the $O(2\,\text{cm})$ resolution is achieved in the high p_T region, but further improvements are necessary in the low p_T region.

2.2 Cosmic ray setup

The first practical test of the z vertex trigger is the upcoming CDC cosmic ray test. Only a part of the CDC is used (two sectors with opposite φ), and there is no magnetic field. Cosmic tracks passing

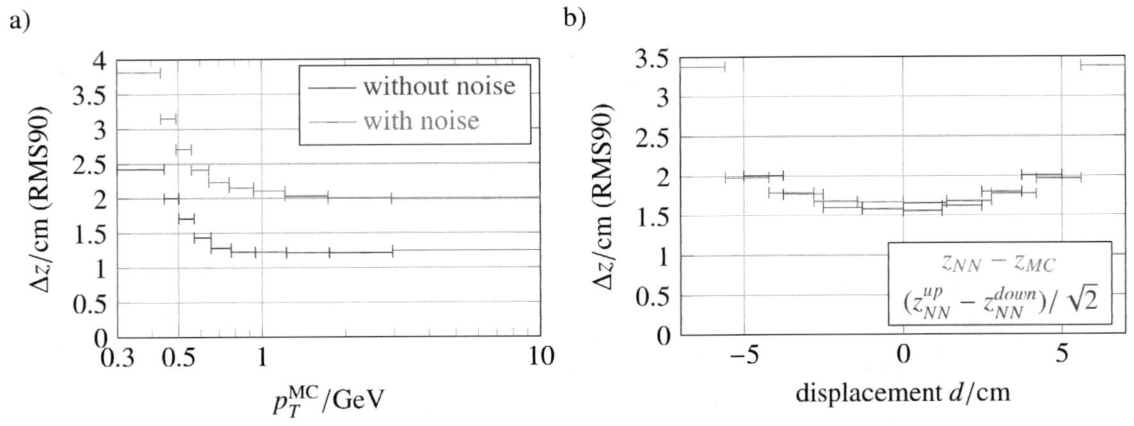

Figure 5. a) z vertex resolution using two neural networks for different charges on ideal noiseless events and realistic events with noise. b) z resolution in the cosmic ray setup. The red line shows single track z vertices vs. simulated values. The blue line shows a comparison of the apparent two track z vertices. The displacement d is the distance of closest approach to the z axis.

close to the z-axis will usually leave hits in both the upper and the lower half of the CDC and should be recognized as two tracks by the 2D finder. Due to the larger distance of closest approach to the z-axis (d) and the 2D finder constraint for tracks to pass through the z axis, displaced tracks will be recognized as curved tracks. Dedicated neural networks are trained with the altered event topology and detector setup. As reference value, the tracks from the upper CDC half are compared to the tracks in the lower CDC half. Figure 5 shows the z vertex resolution of simulated cosmic ray tracks vs. the displacement to the z axis. It shows that the neural network can easily adapt to the altered conditions.

3 3D track finding

A track finding algorithm is crucial to enable the application of the neural networks to single tracks. It uses the CDC hits as input to find the initial track estimates and relate them to the hits. The correct hit selection is essential for the z estimation accuracy, but difficult in the presence of noise. The track estimates are used to calculate the neural network inputs and to enable phase space sectors with "expert" neural networks. The 2D finder in the foreseen setup [1, 11] only uses the axial hits as input and finds 2D track estimates (p_T, φ). In order to allow the z vertex estimation with the neural network, the stereo hits subsequently need to be related to the 2D estimates.

The recently developed 3D track finder is able to use stereo hits and axial hits simultaneously in order to find the 3D track estimates $(p_T, \varphi, \vartheta)$ and their hit relations. The early use of the additional stereo hit information helps to improve the track finding efficiency, the correctness of the stereo hit selection and the accuracy of the track parameter estimates. The sectors around the track estimates used to select the neural network input (one hit per SL per track, see Subsection 2.1) become significantly smaller for the stereo hits. Additionally, a phase space sectorization in ϑ becomes possible. The method is related to the phase space sectorization developed in [5, 6].

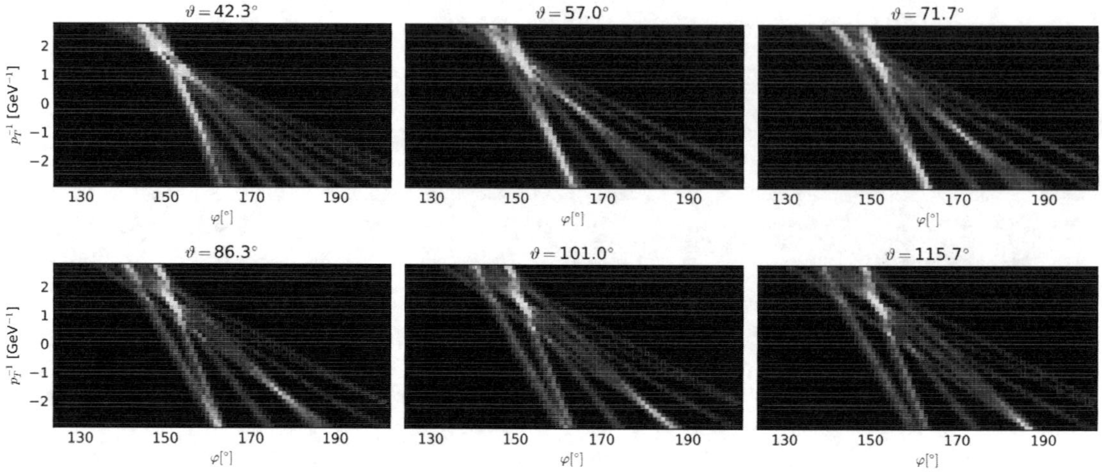

Figure 6. 3D track parameter space for a simulated single muon track with track parameters $p_T^{-1} = 1.62\,\mathrm{GeV}^{-1}$, $\varphi = 151°$, $\vartheta = 38°$ after summing up the hits. For 6 different ϑ bins a 2D track parameter plane in p_T^{-1}, φ is shown. Only close to the correct ϑ, the lines of the stereo hits intersect with the crossing point of the axial hits.

3.1 Track finding setup

The basic concept of the proposed 3D finder is the Bayesian principle of conditional probabilities. A model of the joint probability space of hits and tracks can be formulated as

$$P(T|H) = \frac{P(H|T) \times P(T)}{P(H)}$$

where $P(T|H)$ denotes the probability that a set T of tracks has caused the observed set H of hits. In the proposed approach, the right hand side of the equation is approximated by a multi-dimensional histogram of dimension $\dim(H) + \dim(T)$ that can be trained using simulated events. With this general formulation, independent of the hit and track model, the concrete hit and track parametrization can easily be adapted to varying conditions.

Similar to a Hough transformation [4], the track finding problem is transformed to a peak finding problem in the track parameter space. Peaks can be interpreted as local maxima of the probability distribution, i.e. the tracks most probable to have caused the observed hits. Instead of analytically calculating the single hit representations in track parameter space, they are obtained from a histogram using simulated tracks. A weight $w(t)$ for a specific track t in the track parameter space ($t \in T$), given a set H, is calculated as the sum over all single hit contributions $P(t|h)$ for this track t:

$$w(t) = \sum_{h \in H} P(t|h) \tag{1}$$

In the special case of using only axial hits and ignoring ϑ in the track parametrization, the results are equivalent to the analytically calculated Hough transformation implemented by the current 2D finder. For a specific value of ϑ, a 2D track parameter plane can be constructed from the stereo hits and used in the same way as the 2D plane from the axial hits.

DOI: 10.1051/epjconf/201612700016

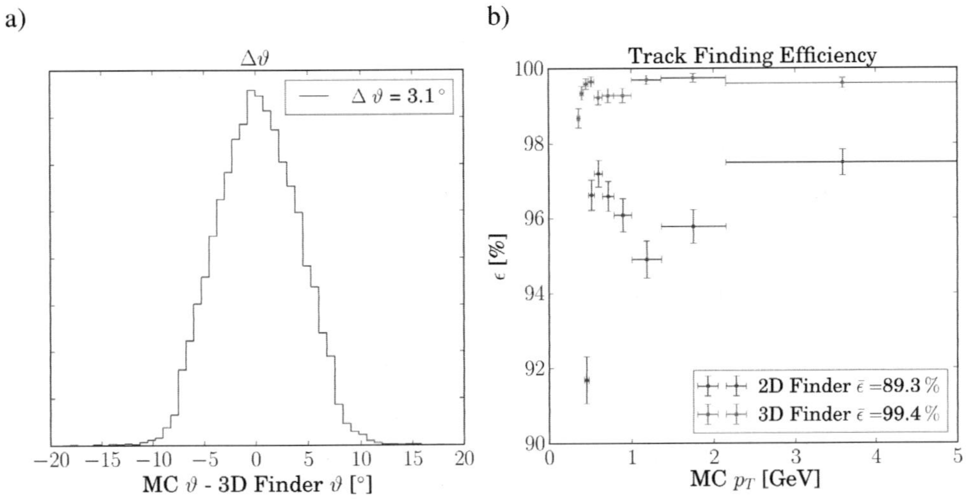

Figure 7. Results of the Bayesian 3D track finder. a) Accuracy of the ϑ estimation. b) Track finding efficiency compared to a 2D finder.

3.2 Training of the track finder

In the proposed 3D finder the distribution $P(t|h)$ is approximated by a 5D histogram representing a 3D track space (p_T, φ, ϑ) and a 2D hit space (TS-id and priority). With 2336 TS and 3 options for the priority, the hit space is already discrete. On the track space the following binning is applied: 384 bins in φ, 40 bins in p_T and 6 bins in ϑ. Drift times and the left/right information are currently unused, but they could easily be included in a future extension of this model. The training is carried out by filling the histogram, i.e. incrementing each (track, hit) pair for a sufficiently large set of training events. To finalize the training, the filled histogram is normalized such that each track is equally probable. During the normalization, the bit width of the histogram is limited to 3 bits.

3.3 Application of the track finder

A track finding and track parameter estimation can be initiated by looking up the 3D track spaces corresponding to each hit in an event and then summing them up (Equation (1)). An example of the summed 3D track space is shown in Figure 6. In the 6 ϑ bins a p_T^{-1}, φ plane is shown, containing one line for each SL. The lines of the five axial SL have the same intersection point in all ϑ bins, whereas the four stereo lines have the same intersection point only in the correct ϑ bin.

The next step is peak finding, which can be done by a clustering algorithm. The current setup requires cluster members to have a weight of at least 90% of the peak weight. Once clusters in track space are found, the contributing hits can be selected. Only hits with a high weight contribution to a cluster are related to the found track. Finally, an estimate of the track parameters for a found cluster is calculated with its related hits as the weighted mean of the cluster in track space. By using the weighted mean, with a 3 bit weight contribution of each hit to each cluster cell, the track parameter resolution can become smaller than the bin width.

The achievable ϑ resolution (RMS 90%) and the track finding efficiency ϵ, with the proposed 3D track finder compared to a 2D finder, are shown in Figure 7. For better illustration only the region with $\epsilon > 90\%$ is shown, whereas, for the 2D finder, the two points at the lowest p_T values are cut (values: $\epsilon = 48.5\%$ and 78.4%). The experiment was carried out with 10000 simulated two-muon events, with a uniform distribution in the track parameters $p_T^{-1}, \varphi, \vartheta$ and ranges $p_T \in [0.35, 5], \varphi \in [0, 360]°, \vartheta \in [35, 123]°$. The good resolution in ϑ of $\Delta\vartheta \approx 3°$ clearly enables the possibility to use 3D input information for the neural network. The efficiency plot illustrates the significant improvement of a 3D finder compared to a 2D finder. The low efficiency of the 2D finder is due to the requirement that hits are present in all five axial SLs. In particular, the dip around 1 GeV in Figure 7 is caused by an inefficiency of the TSF for certain crossing angles. By including the stereo hits, the 3D finder can also identify tracks with missing TS hits, reducing the effect of the TSF inefficiency and at the same time increasing the acceptance in p_T and ϑ.

Compared to the 2D track finder, the 3D track finder requires more hardware resources. It has to receive approximately twice the number of TS as input, the phase space has to be 6 times larger for the ϑ bins, and it requires 3 bits for each bin in $P(t|h)$. The possibilities for a hardware implementation will be evaluated by future studies. If the track finding in 3D is too exhaustive for the present hardware, the method is alternatively suited for the ϑ estimation only. The p_T, φ part of the 3D finder is analog to the present 2D finder, such that the ϑ estimation can be executed as a separate step after the 2D finder.

4 Conclusion

A z vertex trigger is essential for the Belle II experiment in order to cope with the huge beam background levels coming along with the luminosity upgrade. Specifically, the efficiency for low multiplicity decay channels can be significantly enhanced, e.g. by a factor of ≈ 4 for τ-lepton events. With its high accuracy and stability in the presence of noise, the presented neural network algorithm is well suited for the single track z vertex estimation. However, its accuracy is sensitive to the quality of the preprocessing. Hence, further optimization will mostly focus on the improvement of the track finder and the hit selection. The neural network setup is prepared for the first cosmic ray run, where the correctness of the hardware implementation will be evaluated.

A 3D track finder motivated by the Bayesian principle of conditional probabilities has been presented in a machine learning setup. Simulation studies have demonstrated a high accuracy in 3D track parameter estimation and hit selection. Compared to a 2D track finder, a significant improvement in the track finding efficiency has been achieved. Future studies will evaluate the hardware implementation of the proposed 3D track finder.

References

[1] T. Abe et al., *Belle II Technical Design Report* (2010), http://arxiv.org/abs/1011.0352

[2] A. Abashian et al., Nucl. Inst. & Meth. A **479**, 117 (2002)

[3] C. Bernardini et al., Phys. Rev. Lett. **10**, 407 (1963)

[4] P.V.C. Hough, *Machine Analysis Of Bubble Chamber Pictures*, in *HEACC, Proc. 2nd Int. Conf. on High-Energy Accelerators and Instrumentation* (1959), Vol. C590914, pp. 554–558, http://inspirehep.net/record/919922/files/HEACC59_598-602.pdf

[5] S. Skambraks, Diplomarbeit, Institute for Informatics, Ludwig-Maximilians-Universität, München (2013), https://publications.mppmu.mpg.de/2013/MPP-2013-339/FullText.pdf

[6] S. Skambraks, M.S. thesis, Faculty of Physics, Ludwig-Maximilians-Universität, München (2013), https://publications.mppmu.mpg.de/2013/MPP-2013-377/FullText.pdf

[7] S. Skambraks et al., IEEE Trans. Nucl. Sci., **62**, 1732 (2015)

[8] S. Neuhaus et al., J. Phys. Conf. Ser. **608**, 012052 (2015)
[9] R. Hecht-Nielsen, *Theory of the backpropagation neural network*, in *IJCNN, IEEE Int. Joint Conf. Neural Networks* (1989), Vol. 1, pp. 593–605
[10] C. Igel, M. Hüsken, *Improving the Rprop Learning Algorithm*, in *Proc. 2nd Int. Symp. Neural Computation* (2000), pp. 115–121
[11] Y. Iwasaki et al., IEEE Trans. Nucl. Sci. **58**, 1807 (2011)

EPJ Web of Conferences **127**, 00017 (2016) DOI: 10.1051/epjconf/201612700017
Connecting The Dots 2016

FPGA-Based Approach to Level-1 Track Finding at CMS for the HL-LHC

Louise **Skinnari**[1,a] for the CMS Collaboration

[1] *Cornell University, Ithaca, NY, USA*

Abstract. The high luminosity upgrade of the LHC is expected to deliver luminosities of $7.5 \times 10^{34}\,\mathrm{cm}^{-2}\mathrm{s}^{-1}$, with an average of 140–200 overlapping proton-proton collisions in each bunch crossing at a frequency of 40 MHz. To maintain manageable trigger rates under these conditions track reconstruction will be incorporated in the all-hardware first level of the CMS trigger. A track-finding algorithm based on seed tracklets has been developed and implemented on commercially available FPGAs for this purpose. An overview of the algorithm is presented, results are shown of its expected performance from simulations, and an implementation of the algorithm in a Xilinx Virtex-7 FPGA for a hardware demonstrator system is discussed.

1 Introduction

The upgrade to the High Luminosity LHC (HL-LHC) is planned for 2024–2025 and is expected to deliver a peak luminosity of about $7.5 \times 10^{34}\,\mathrm{cm}^{-2}\mathrm{s}^{-1}$ with the goal of delivering $3000\,\mathrm{fb}^{-1}$ per experiment over a ten-year period [1]. The high instantaneous luminosity is predicted to yield an average number of overlapping proton-proton collisions per bunch crossing (pileup) of between 140 and 200. These extreme pileup (PU) conditions create a challenging environment for the LHC experiments and place stringent requirements on the detectors' trigger systems. The CMS [2] physics program for the HL-LHC requires maintaining manageable trigger rates with low thresholds similar to those achieved during Run 1. An integral part of achieving this goal is the addition of track reconstruction in the all-hardware Level-1 (L1) trigger. A 40 MHz silicon-based track trigger on the scale of the CMS detector has never been built and it is therefore essential to demonstrate the feasibility of such a system.

The tracklet approach for track reconstruction at L1 is a road-search algorithm, implemented using commercially available field-programmable gate array (FPGA) technology. The ever-increasing capability and the programming flexibility make FPGAs ideal for performing fast track finding. The tracklet approach allows a naturally pipelined implementation with low time multiplexing (typically a factor of 4–8). It also allows for a simple emulation of the algorithm for the full detector.

These proceedings discuss the motivation for incorporating tracking at L1 and give an overview of the CMS track-trigger concept. The tracklet algorithm is described in detail and results of the estimated performance of the algorithm, based on simulation studies in C++ and Verilog, are presented. The development of a hardware demonstrator system using Xilinx Virtex-7 FPGAs is also discussed.

[a] e-mail: louise.skinnari@cern.ch

2 Motivation for L1 tracking

At the HL-LHC, L1 trigger rates for single muons, electrons, and jets will exceed the current front-end capabilities. Significantly increasing the trigger thresholds would limit the physics potential and would alone be insufficient. The addition of a L1 track trigger is an integral part of maintaining the physics acceptance at the HL-LHC. Including tracking in the L1 trigger allows for improved lepton identification and momentum measurements, the addition of track isolation, and performing vertex identification for hadronic triggers. These additional handles yield large potential reductions of the L1 trigger rates. One example is illustrated in Figure 1, which shows the efficiency for a single muon trigger with $p_T > 20\,\text{GeV}/c$ as a function of the simulated muon p_T (left) and the trigger rate as a function of the muon p_T threshold (right). The red curves show the efficiencies and rates with the stand-alone L1 muon trigger system for $|\eta| < 1.1$ (filled circles) and $1.1 \leq |\eta| \leq 2.4$ (open boxes). The black curves show the efficiencies and rates when incorporating tracking at L1. With the stand-alone muon trigger, low-p_T muons can be misassigned as a high-p_T muon, causing a flattening of the trigger rate with increasing p_T threshold. Using L1 tracking yields an improved momentum measurement and a sharp turn-on at the threshold, which in turn translates to a reduced trigger rate [3].

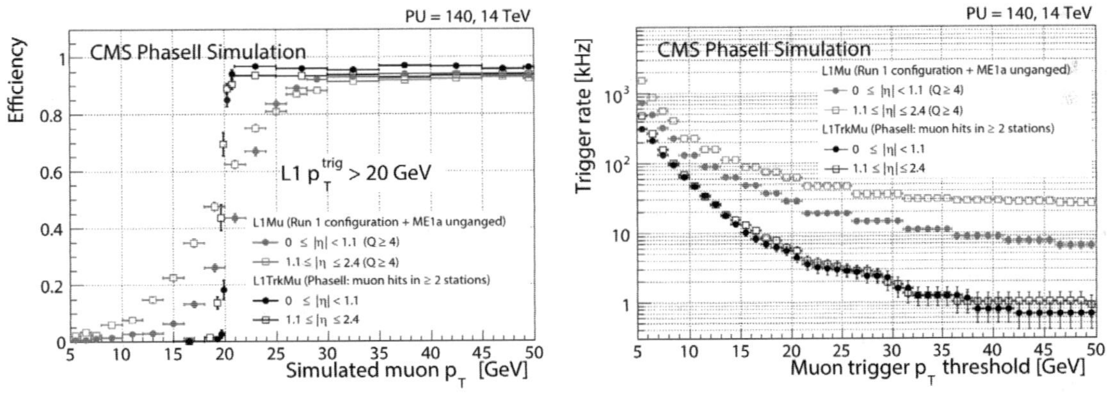

Figure 1: The efficiency of a single muon trigger with $p_T > 20\,\text{GeV}/c$ as a function of the simulated muon p_T (left) and the trigger rate as a function of the muon trigger threshold (right), shown for the stand-alone muon trigger (red) and when including L1 tracking (black) for $|\eta| < 1.1$ (filled circles) and $1.1 \leq |\eta| \leq 2.4$ (open boxes) [3].

3 Track trigger concept

The upgraded all-silicon tracker consists of an outer tracker with a central barrel and two endcaps as well as an inner pixel detector, which is not part of the L1 track trigger. Unique to the new outer tracker is the use of "p_T modules" that provide p_T discrimination already in the front-end readout electronics through hit correlations between two closely spaced sensors. Two different module types are employed: pixel-strip (PS) modules in the inner barrel layers and inner half of the disks, and strip-strip (2S) modules in the outer barrel layers and outer half of the disks [3]. The 2S modules are composed of two superimposed strip sensors of about $10 \times 10\,\text{cm}^2$, mounted with the strips parallel to one another. The PS modules have two sensors of about $5 \times 10\,\text{cm}^2$, where one sensor is segmented in strips and the other is composed of $10\,\mu\text{m} \times 1.5\,\text{mm}$ "macro-pixels". The pixelated sensor provides a precise measurement of the z coordinate, which in turn enables primary vertex discrimination at L1. The output from the PS and 2S modules are correlated pairs of clusters, referred to as "stubs", that are required to be consistent with a $p_T > 2\,\text{GeV}/c$ track. The stubs are the input to the back-end processing system that performs the L1 track finding. Since only ~5% of tracks in minimum-bias events have $p_T > 2\,\text{GeV}/c$, forming stubs results in a significant rate reduction already for the L1 track input data.

4 Tracklet approach

The goal of the L1 track finding is to reconstruct the trajectories of charged particles with $p_T > 2\,\text{GeV}/c$ and to identify the track z_0 position (the z coordinate where the track intercepts the z-axis) with about 1 mm precision, similar to the average vertex separation in events with an average pileup of 140. The proposed tracklet method forms track seeds (tracklets) from pairs of stubs in adjacent layers or disks of the outer tracker. The tracklets provide roads where compatible stubs are included to form track candidates. A linearized χ^2 fit determines the final track parameters.

The tracklet algorithm has been optimized to provide full coverage of the tracker with the least amount of redundancy in data duplication. The system is divided into 28 sectors in the r-ϕ-plane, where the processing for each sector is done by a single FPGA. This division limits the need for board-to-board communication to the nearest neighbor on each side.

4.1 Algorithm overview

The tracklet algorithm begins with a seeding step where tracklets are formed from pairs of stubs in adjacent layers or disks. An initial estimate of the tracklet parameters is calculated from the two stubs and using the detector origin as a constraint in the r-ϕ-plane. A tracklet must be consistent with having $p_T > 2\,\text{GeV}/c$ and $|z_0| < 15\,\text{cm}$. The seeding is performed multiple times to ensure good coverage for the full η range and for redundancy in the system. The current implementation performs seeding between barrel layers 1+2, 3+4, and 5+6, between disk 1+2 and 3+4, and between barrel layer 1+disk 1. The tracking efficiency for different seeding combinations is shown for single muons in Figure 2, using an integer-based C++ emulation of the algorithm as it would be implemented on a FPGA.

The tracklets are projected to other layers and disks to search for matching stubs. The projections use predetermined search windows, derived from residuals between projected tracklets and stubs. The tracklets are projected both inside-out and outside-in. If a matching stub is found, the stub is included in the track candidate and the difference between the projected tracklet and the stub position is stored. A linearized χ^2 fit is performed for stubs matched to the trajectory. The track fit implementation uses precalculated derivatives and the tracklet–stub position differences from the projection step. The linearized χ^2 fit corrects the initial tracklet parameters to give the final track parameters p_T, η, ϕ_0, z_0 (and optionally d_0).

A given track can be found many times due to seeding in multiple pairs of layers. Duplicate tracks are removed by comparing tracks in pairs, counting the number of independent and shared stubs. Figure 3 shows for single-muon events the number of tracks found per event prior to any duplicate removal, after duplicate removal between tracks within a ϕ sector, and after removing duplicates also between neighboring sectors.

4.2 Performance studies

The estimated performance of the tracklet algorithm is studied with a C++ simulation of the algorithm using floating-point precision in the calculations. The L1 tracking efficiency as a function of η and

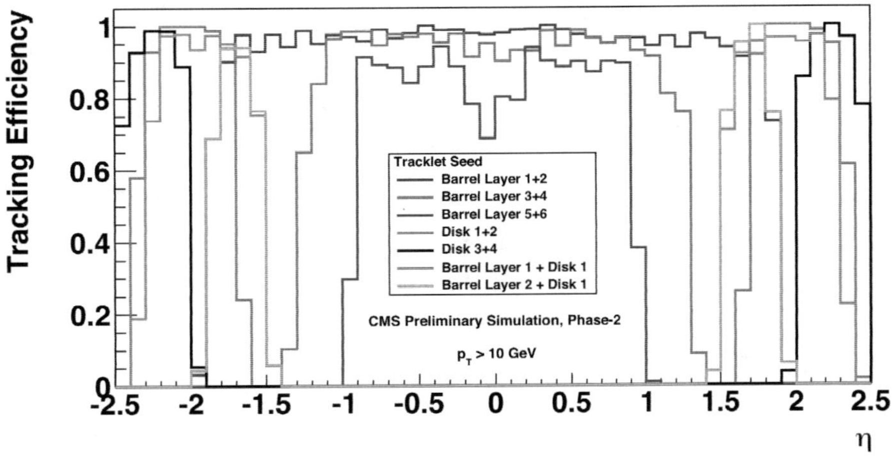

Figure 2: Tracking efficiency for single muons with $p_T > 10\,\text{GeV}/c$ in events without pileup, shown as a function of muon η for different layer and disk combinations in forming the seed tracklets.

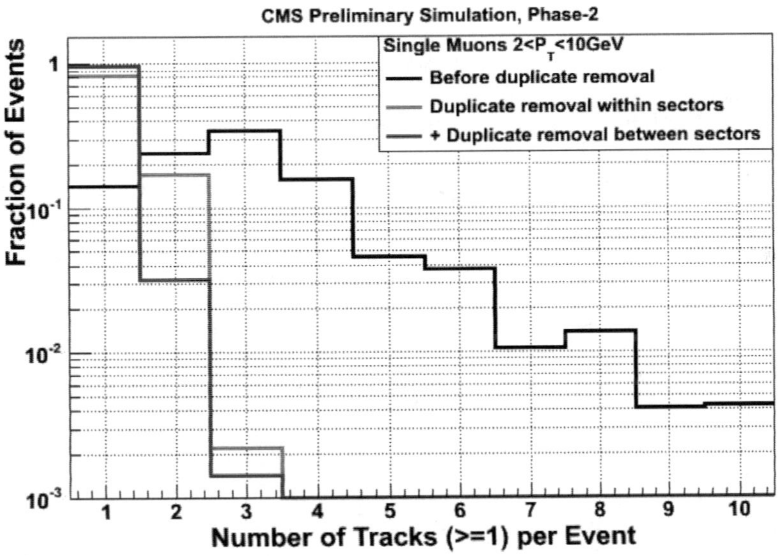

Figure 3: Number of tracks per event for single muons with $2 < p_T < 10\,\text{GeV}/c$ in events without pileup. The number of tracks are shown before any duplicate removal (black) and after duplicate removal within sectors (red), or after also removing duplicates between neighboring sectors (blue).

DOI: 10.1051/epjconf/201612700017

p_T is shown in Figure 4 for muons, pions, and electrons in events with <PU>=140. For $|\eta| < 1.0$ and $p_T > 2\,\mathrm{GeV}/c$, an efficiency of >99%, 90%, and 89% is observed for muons, pions, and electrons, respectively. Pions and electrons have a lower efficiency compared to muons due to their additional interactions with the detector material.

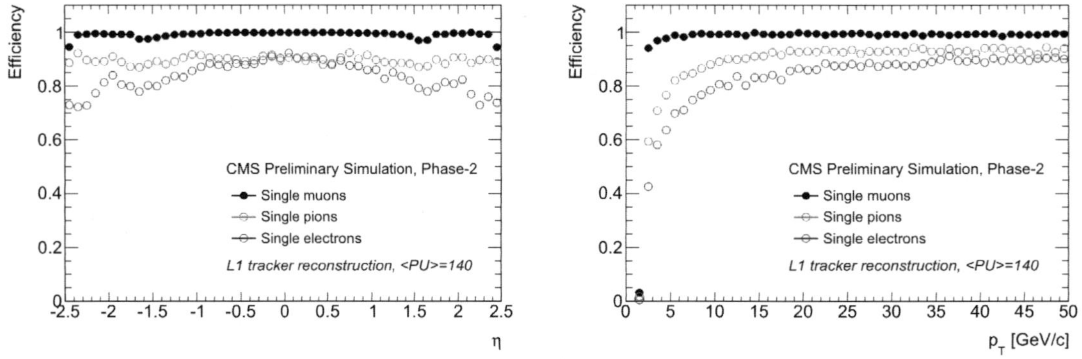

Figure 4: Efficiency for L1 track reconstruction as a function of η (left) and p_T (right) for muons, pions, and electrons in events with <PU>=140 [3].

The L1 track z_0 resolution and the relative resolution in p_T are shown in Figure 5 for single muons as a function of $|\eta|$ for three different ranges of p_T. The z_0 resolution is about 1 mm, similar to the average separation of pileup vertices, for a wide range of $|\eta|$. The momentum resolution is about 1% for high-p_T tracks for $|\eta| < 1.5$. For higher values of $|\eta|$ the momentum resolution decreases due to the increased extrapolation distance to the beam axis, where the track parameters are calculated. The precise z_0 resolution allows the selection of tracks originating from a common vertex for use in L1 trigger algorithms and the accurate p_T resolution results in sharp muon trigger thresholds.

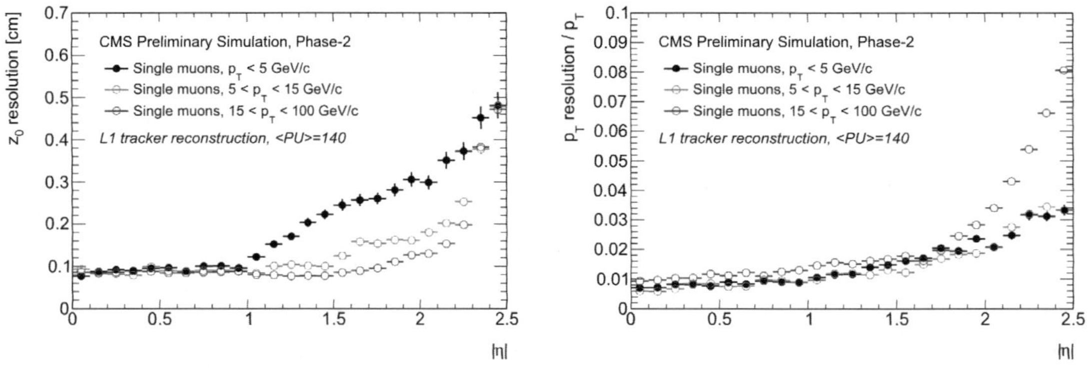

Figure 5: Resolution in z_0 and relative resolution in p_T for the L1 track reconstruction of single muons in events with <PU>=140, shown as a function of $|\eta|$ for different p_T ranges [3].

The tracklet algorithm has been fully implemented both using floating-point precision in the calculations, as shown above, and as an integer-based C++ emulation of the algorithm as it would be implemented on a FPGA. The floating-point simulation and the integer-based emulation are in good agreement. As an example, track parameter resolutions are compared between the floating-point simulation and the integer-based emulation in Figure 6 for muons with $5 < p_{\mathrm{T}} < 10\,\mathrm{GeV}/c$ and $1.2 < |\eta| < 2.4$. The root mean square (RMS) of each distribution is calculated. The resolutions as obtained with the integer-based emulation and the (exact) floating-point simulation are compared using $(\mathrm{RMS}_{\mathrm{integer}} - \mathrm{RMS}_{\mathrm{float}})/\mathrm{RMS}_{\mathrm{float}}$, calculated to be 0.08, 0.10, 0.12, and 0.13 for the relative p_{T}, ϕ, η, and z_0 resolution, respectively, demonstrating a good level of agreement.

5 Hardware system

The primary challenges in building a L1 track trigger system are the large data throughput from the detector at 40 MHz and the short time available to reach a trigger decision. The upgraded L1 trigger will have an overall latency, i.e. the maximum time during which events can be stored in on-detector pipeline memories, of 12.5 μs, where 4 μs are allocated to performing the L1 track finding.

To address these challenges, the tracklet hardware configuration relies on parallel data processing. The detector is divided into sectors in the r-ϕ-plane. The current baseline uses 28 ϕ sectors, so that tracks with $p_{\mathrm{T}} > 2\,\mathrm{GeV}/c$ span maximally two sectors. This limits the need for data transfer between sectors to the nearest neighbor on each side. A dedicated processing board is used for each sector.

To increase the time available for data processing, the system is time multiplexed. The current baseline assumes a time-multiplex factor of six, so that each copy of the system receives a new event every 150 ns for a 40 MHz bunch crossing frequency. A different time-multiplex factor of e.g. four or eight (receiving a new event every 100 or 200 ns) could also be considered in optimizing for latency or available processing time. The tracklet formation is performed within sectors and a small amount of data is duplicated in every other layer to avoid gaps. Tracklets that project to a neighboring sector are sent there for tracklet–stub matching. By construction, the system operates with a fixed latency.

5.1 Firmware implementation

The tracklet algorithm is implemented in firmware as eight processing steps and two transmission steps [4]:

- *Stub organization:* (1) Sort the input stubs by their corresponding layer, and (2) into smaller units in z and ϕ, forming what are referred to as "virtual modules".
- *Tracklet formation:* (3) Select possible tracklets from allowed stub pairs using lookup tables, and (4) calculate the initial tracklet parameters.
- *Projections:* (5) Transmission of projections pointing to neighboring sectors. (6) Route the projections based on smaller units in z and ϕ.
- *Stub matching:* (7) Match projected tracklets to stubs, and (8) calculate the difference in position between the stubs and projected tracklet. (9) Transmission of matches between sectors.
- *Track fit:* (10) Perform track fit, correcting the initial tracklet parameter estimate.

With a time-multiplex factor of six, a new event is received every 150 ns at which point the previous event moves to the following processing step. The number of operations performed at each step, such as the number of stub pairs that can be tried in the formation of a valid tracklet, will depend on the clock speed at which the project can run with the hardware used.

EPJ Web of Conferences **127**, 00017 (2016) DOI: 10.1051/epjconf/201612700017
Connecting The Dots 2016

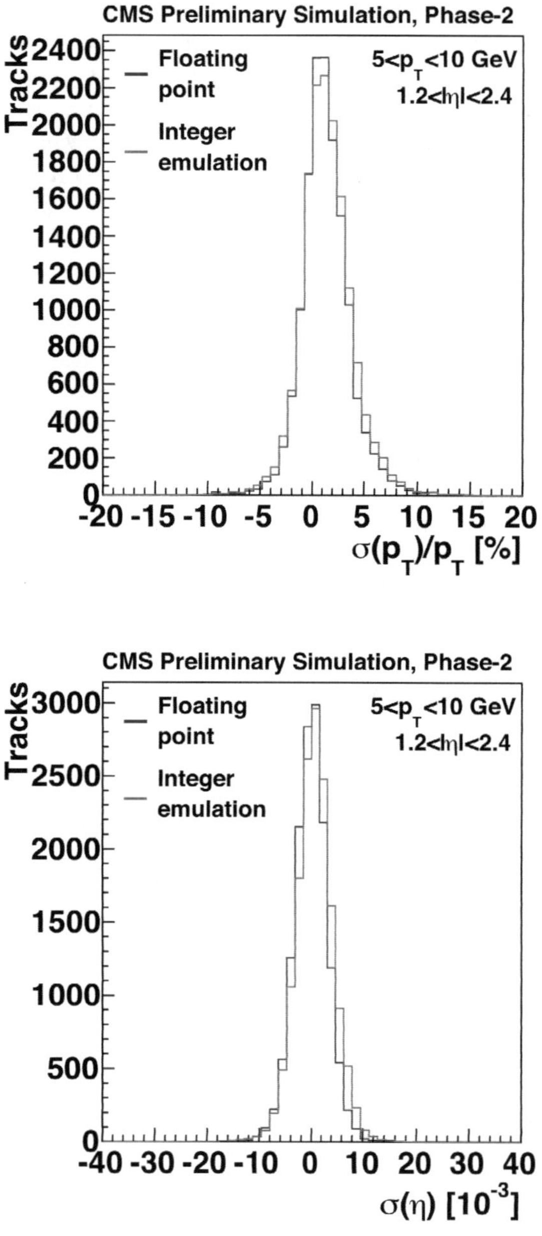

Figure 6: Relative resolution in p_T (top left), resolution in ϕ_0 (top right), η (bottom left), and z_0 (bottom right), for muons with $5 < p_T < 10\,\mathrm{GeV}/c$ and $1.2 < |\eta| < 2.4$ in events without pileup.

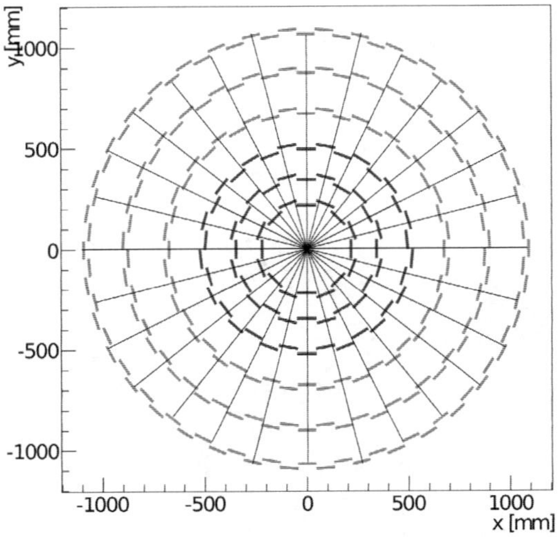

Figure 7: Schematic view in the r-ϕ-plane of the upgraded outer tracker, dividing into 28 ϕ sectors so that a track with $p_T > 2\,\text{GeV}/c$ spans maximally two sectors, limiting the communication between sectors. PS modules (blue) and 2S modules (red) are used in the three inner and outer layers, respectively.

The combinatorics in forming tracklets and matching projections to stubs is efficiently reduced by dividing sectors into smaller units in z and ϕ to allow additional parallel processing. These smaller units are, as mentioned above, referred to as "virtual modules". Only a small fraction of virtual module pairs can form a valid tracklet – the majority would be inconsistent with a track with $p_T > 2\,\text{GeV}/c$ and $|z_0| < 15\,\text{cm}$. This subdivision efficiently reduces the number of combinations that need to be tried by the algorithm already before doing any event processing.

As the system operates at fixed latency, a given processing step can perform a fixed number of operations. Initial studies of the impact of truncation, such as if there are more stub pairs to try than can be processed within the time constraints of the tracklet formation step, show a robustness of the algorithm to these effects. This is achieved through the use of virtual modules, and the redundancy in using multiple seeding pairs.

The tracklet algorithm including the track fit has been implemented in firmware for half the barrel ($z > 0$) for one sector in ϕ. Work is ongoing to expand the project to include the forward disks. The tracklet algorithm in the disks is already fully implemented as an integer-based emulation. While the duplicate removal algorithm has not yet been incorporated into the main tracklet project, it is fully implemented in the floating-point simulation, the integer-based emulation, and as a standalone algorithm in firmware on a VC709 evaluation board with a Virtex-7 FPGA [5].

5.2 Hardware demonstrator

A system hardware demonstrator has been set up and is now being used for testing the tracklet project. The goal of the demonstrator is to show that the full tracking chain meets the required performance within the available latency. The demonstrator includes three ϕ sectors and one time-multiplexing

 DOI: 10.1051/epjconf/201612700017

slice. A total of four boards are used, one for the central ϕ sector, two for its nearest neighbors, and one board that is shared for sending the stub input and receiving the track output.

The boards used for the demonstrator system are μTCA boards with a Xilinx Virtex-7 (XC7VX690T) FPGA [5] and a Xilinx Zynq processor for outside communication. These so-called CTP7 boards [6] were developed for the current CMS L1 trigger upgrade [7]. An AMC13 [8] card provides the central clock distribution. The inter-board communication uses 8b/10b encoding with 10 Gbit/s link speed. The demonstrator system and a CTP7 board are shown in Figure 8.

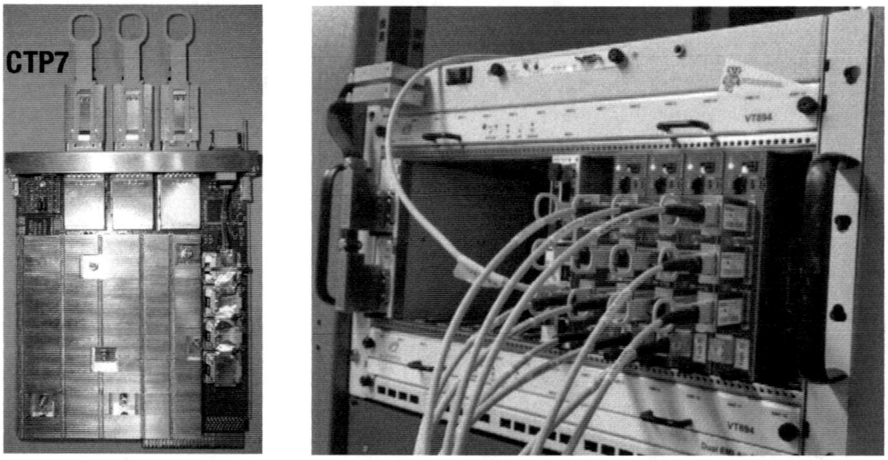

Figure 8: A CTP7 board (left) and the tracklet demonstrator system (right).

One ϕ sector for a half-barrel has been implemented on the CTP7 boards, running at 240 MHz. Events are processed through the demonstrator as illustrated in Figure 9. First, input stubs obtained from simulations are written to the data trigger and control (DTC) emulator board. On a GO signal, stubs are sent to the three sector processor boards. A new event is sent to each sector board every 150 ns. The events are processed through the sector boards and projections and matches are sent to and received from neighboring boards as required. The final output tracks are received by the track sink board. Systematic studies are performed to compare the integer-based emulation of the tracklet algorithm with a simulation of the firmware using Vivado, as well as with the output tracks from the demonstrator system. Full agreement is observed in processing single-track events between the emulation, firmware simulation, and board output. Studies of many-track events with high pileup are ongoing.

6 Conclusions

CMS will require a new all-silicon tracking system with triggering capabilities for the HL-LHC in order to maintain high trigger efficiencies and manageable rates without large increases in trigger thresholds. The tracklet approach is one of the proposed methods for performing L1 track finding. The method is based on a road-search algorithm and uses commercially available FPGA technology. The tracklet algorithm is implemented both as a floating-point simulation and an integer-based emulation of the algorithm as implemented in an FPGA. Half of the barrel has been implemented for a sector in ϕ in a Virtex-7 FPGA. The final system, which will not be built for many years, will use future-generation (more powerful) FPGAs. To demonstrate the system feasibility, a hardware demonstrator

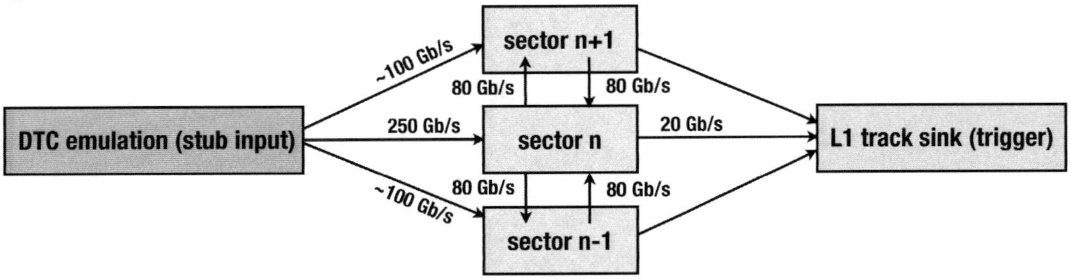

Figure 9: Schematic overview of the tracklet demonstrator system. The DTC emulator and L1 track sink use a single board.

system based on Virtex-7 FPGAs has been assembled and is currently being used to validate the algorithm and board-to-board communication, to measure timing and latency, and to establish the algorithm performance. Preliminary studies from the demonstrator, processing events from the input stubs to the final output tracks, show promising results for the feasibility of the system within the available latency.

Acknowledgements

This work was supported by the US National Science Foundation through the grant NSF-PHY-1307256 . I also wish to thank my tracklet colleagues at Cornell, Rutgers, Ohio State, Notre Dame, and UC Riverside.

References

[1] F. Zimmermann, PoS **EPS-HEP 2009**, 140 (2009), https://cds.cern.ch/record/1211584
[2] The CMS Collaboration, JINST **3**, S08004 (2008)
[3] The CMS Collaboration, Tech. Rep. CERN-LHCC-2015-010, CERN, Geneva, Switzerland (2015), https://cds.cern.ch/record/2020886
[4] J. Chaves, JINST **9**, C10038 (2014)
[5] Xilinx Incorporated, *7 Series FPGAs Overview*, last accessed on 2016-07-21, pdf
[6] A. Svetek et al., JINST **11**, C02011 (2016)
[7] The CMS Collaboration, Tech. Rep. CERN-LHCC-2013-011, CERN, Geneva, Switzerland (2013), https://cds.cern.ch/record/1556311
[8] Boston University, *The AMC13 Project*, last accessed on 2016-07-21, pdf

AUTHOR INDEX